The Feejee Mermaid

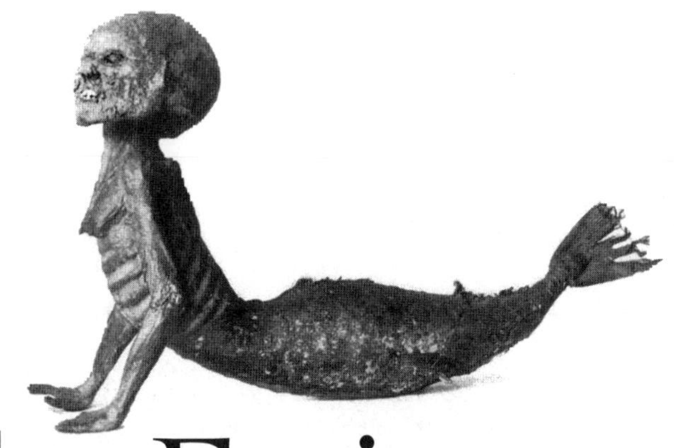

The Feejee Mermaid

and Other Essays in Natural and Unnatural History

Jan Bondeson

CORNELL UNIVERSITY PRESS
Ithaca and London

First published 1999 by Cornell University Press

Printed in the United States of America

Library of Congress Cataloging–in–Publication Data

Bondeson, Jan.
The feejee mermaid and other essays in natural and unnatural history / Jan Bondeson.
p. cm.
ISBN 0-8014-3609-5 (cloth : alk. paper)
1. Zoology—Miscellanea. 2. Zoology—Anecdotes. I. Title.
QL58.B57 1999
590—dc21 98–38295

Cornell University Press strives to use environmentally responsible suppliers and materials to the fullest extent possible in the publishing of its books. Such materials include vegetable-based, low-VOC inks and acid-free papers that are recycled, totally chlorine-free, or partly composed of nonwood fibers.

Cloth printing 10 9 8 7 6 5 4 3 2 1

Contents

A Prelude

In my book collection, there is a shelf for eighteenth- and nineteenth-century natural history books intended for the general public. Some of these, such as Oliver Goldsmith's *History of the Earth and Animated Nature,* and two cheap editions of Buffon's *Natural History,* are encyclopedias intended to present a complete overview of the animal kingdom. Others, such as Sir Ray Lankester's *Diversions of a Naturalist* and Philip Henry Gosse's excellent *Romance of Natural History,* are collections of essays about various remarkable occurrences and unsolved mysteries concerning the animal kingdom. Intended to amuse as well as to educate, these books were among the best sellers of their time; within their particular genre, they were eclipsed only by Frank Buckland's *Curiosities of Natural History,* a justly famous collection of essays on all kinds of oddities and peculiarities within the natural sciences.

Buckland was a qualified surgeon, but his passion was for practical zoology. He soon exchanged his post as assistant surgeon to the Second Life Guards for a more congenial job as inspector of fisheries. A copious and amusing writer and journalist, he did much to enliven various nineteenth-century periodicals dedicated to hunting, fishing, and animal lore. He was also something of an inventor, and devised many curious schemes, such as making gloves and shoes from the skins of rats. Another of Frank Buckland's interests was the acclimatization of foreign animals into the British Isles: he envisioned herds of eland and antelopes galloping through the Surrey countryside. This was not wholly due to an enthusiasm for zoology, nor was it entirely altruistic, since Frank Buckland had an almost fanatical desire to taste all kinds of animal meat, cooked or uncooked. At his dinner parties, guests were served emu, hippopotamus, elephant's trunk soup, and boiled boa constrictor, the gastronomic merits or demerits of which were discussed at length in his periodical *Land and Water.* A roasted panther was pronounced "not very good," but this is not surprising, as it had actually been dug up by Frank Buckland and his friends after being buried for several days! Through contacts at various zoos and menageries, Buckland could usually procure beef-steaks from various rare animals; when these were in short supply, he had to rely on various household pests, served up as boiled mole, toasted earwigs, ragout of bluebottle flies, or garden slugs masquerading as escargots. Once, at a dinner party, Frank Buckland was observed to have a particularly odd object on his plate. Before consuming it, he announced that "I have eaten many strange things in my lifetime, but never before have I eaten the heart of a king." This was the unceremonious end of the embalmed heart of King Louis XIV of France, originally stolen by grave robbers at St. Denis and sold to Lord Harcourt, a friend of the Buckland family.

This collection of essays on various odd, uncanny, or macabre aspects of the animal kingdom and its relation with humankind is deliberately laid out in a way resembling these nineteenth-century natural history books. As in Frank Buckland's *Curiosities of Natural History,* the reader will encounter learned pigs, sagacious elephants, and dried mermaids on exhibition in London. One of the last was actually purchased by Buckland for his own collection; whether it appeared on his dinner table is a matter for speculation. As in Philip Henry Gosse's *Romance of Natural History,* the reader will encounter little fish raining down from the sky, and toads living for thousands of years immured in blocks of stone.

My aim has been, however, to make the book less whimsical and anecdotal. Like Frank Buckland, I am a doctor, although it would not be bragging to say that I have taken my medical education and career rather more seriously than he did. A rheumatologist by profession, I am currently striving to devise novel ways to treat, and hopefully cure, various inflammatory diseases, primarily rheumatoid arthritis. In particular, we are trying to elucidate the role of various transcription factors (nuclear proteins important for the regulation of various genes) in chronic inflammatory diseases, with the aim to develop gene therapeutic strategies. As could be expected, this novel, far-reaching research project is in character rather like charting a huge cobweb of intracellular signaling pathways by tracing each thread of the gigantic web—a cobweb that is daily increasing in size, owing to the discovery of new enzymes, cytokines, and transcription factors. As a recreation, I have, ever since I was a medical student in the early 1980s, investigated various odd, macabre phenomena in medicine and natural history that are ignored by the modern, rationalist textbooks of the history of science. It is fascinating to pore over old Latin tomes at the British Library, and even more so to look up original manuscripts and illustrations in various dusty and long-forgotten archives. In spite of the pressures of my professional life, my spare time has been devoted to investigating the scientific oddities, anomalies, and mysteries of old — many of them unsolved to this day.

Being fortunate enough to live in London, I have had access to the matchless collections of the British Library, containing not only printed books and pamphlets, but also newspapers, manuscripts, prints and drawings. The zoological folios of the seventeenth century, written by Conradus Gesner, Ulysses Aldrovandi, Edward Topsell, John Jonstone, and others, were obvious sources, and gave useful references to earlier publications. Often, it proved necessary to consult the original sources, and again London's wealth of museums and archives made this possible: the repositories of the Theatre Museum, the Royal College of Surgeons, Natural History Museum, Guildhall Library, Newspaper Library, Museum of Garden History, and many others, have provided invaluable illustrations and source material. To bring the research up to date, I have made extensive use of the modern computerized citation indexes of science and the arts and humanities, among other online bibliographical tools. In many instances, I have attempted not only to deal with zoology and history of science but also to bring in aspects of ethnology, philosophy, and literature. The detection of numerous, and often striking, examples of poetry inspired by the weird phenomena under discussion was due less to deep erudition and years of painstaking study, than to familiarity with the English Poetry Complete database; but perhaps one should not give away all the tricks of the trade.

❧ One of these ancient, mysterious anomalies concerns downfalls of living animals with rain: a shower of high-flying fish, or a steady drizzle of little frogs pattering against the umbrella when one is waiting for the bus outside the South Kensington tube. In the seventeenth and eighteenth centuries, this phenomenon was taken perfectly seriously; cases were reported in the foremost scholarly periodicals and were discussed by the leading zoologists of the time. The *Mirror* of August 4, 1838, was headlined "A Shower of Frogs in London!" In the early 1850s, clouds of frogs again covered the London skyline: when the zoologist Edward Jesse took a stroll in his walled Fulham garden just after a rainstorm, the garden was full of little frogs jumping merrily about. The soil was dry gravel, with no moist spot in which the animals could deposit their spawn; indeed, he had to conclude that they could have arrived there only through the house — or from the sky. There are numerous other instances of showers of frogs, toads, and fish; in many of them, the animals were actually seen to descend with the rain. In spite of the near total silence of present-day zoologists and meteorologists about this impossible or "damned" phenomenon, it is possible to build up a case to support the existence of downfalls of fish, amphibians, and other smaller animals with rain. An even more obscure phenomenon, famous in Victorian time but forgotten today, is the toad in the hole: the notion that toads could live for hundreds, nay thousands, of years while totally enclosed within blocks of stone. This is an anomaly in the true sense of the word: the phenomenon is not only irrational but completely inexplicable. Although it is possible to make some headway concerning the mechanisms involved in the "creation" of a toad in the hole, the matter must be declared *sub judice.*

In 1822, a mermaid was apprehended by customs officials at the East India Docks and imprisoned for several weeks, while her immigration status was being debated. Once at liberty in the metropolis, the mermaid enjoyed two hectic seasons as a celebrated star in show business, before dazzling the Americans on her extended tour of the United States, under the management of P. T. Barnum. This star was the celebrated Feejee Mermaid, who is depicted on the cover of this book. Her unparalleled career has been the subject of much speculation; it appears as if this extraordinary creature was even a prime murder suspect in an episode of the popular television series *The X-Files.* Through the discovery of an early description and illustration of the Feejee Mermaid in the archives of the Royal College of Surgeons of England, it has been possible to demonstrate exactly *what* she was.

In the year 1601, Marocco the dancing horse stood on top of St. Paul's steeple, looking disdainfully down on Mr. Holden's unwieldy old camel prancing about on London Bridge. One hundred and eighty-four years later, another four-legged London celebrity, the Learned Pig, received visitors at Charing Cross. On the night of Saturday, March 4, 1826, the entire intestinal tract of an elephant was clandestinely dumped into the Thames from Waterloo Bridge; the next day, a caravan of carts removed a prodigious amount of well-seasoned elephant beef and grease to the London cat's meat factories. This was the unceremonious end of Chunee, a famous elephant who resided at the unique indoor menagerie at Exeter Change, just off the Strand. For 14 years, Chunee was the favorite of all Londoners. After the 5-ton elephant had run amok in central London, 152 projectiles were needed to kill the infuriated beast. The skeleton alone remained, and was exhibited in the battered den as a memorial to the martyred elephant.

Another, more dangerous beast on show in London was Mr. Salgado's basilisk. This fabulous beast was capable of killing people merely by looking at them, but the Londoners seem to have survived its baleful gaze. Less sinister, but equally curious, was the barnacle tree, or tree-bearing geese, exhibited at Spring Gardens in 1807. These geese growing on trees were related to the marvelous Vegetable Lamb of Tartary, which grew on a stem arising from a large plant. I have paid my regards to both the vegetable lambs residing in London at the present time; one of these is at the Natural History Museum in South Kensington, the other at the Museum of Garden History at Lambeth Palace. It is a fitting tribute to this parliament of wonders—porcine performers, mermaids on show, strange beasts, odd happenings, and impossible events—to quote the description of Bartholomew Fair in William Wordsworth's *The Prelude:*

> *All moveables of wonder, from all parts,*
> *Are here—Albinos, painted Indians, Dwarfs,*
> *The Horse of Knowledge, and the learned Pig,*
> *The Stone-eater, the Man that swallows fire,*
> *Giants, Ventriloquists, the Invisible Girl,*
> *The Bust that speaks and moves its goggling eyes,*
> *All out-of-the-way, far-fetch'd, perverted things*
> *All Freaks of Nature, all Promethean thoughts*
> *Of Man, his dulness, madness, and their feats*
> *All jumbled up together to make up*
> *A Parliament of Monsters . . .*

❧ Mr. A. H. Saxon, Ph.D., of Bridgeport, is thanked for important collaboration in the sections on the Feejee Mermaid, Jumbo, and Chunee, and for a critical reading of the entire manuscript. Professor A. E. W. Miles, London, also read part of the manuscript and provided helpful comments. The Crafoord Foundation (Lund, Sweden) is thanked for several valuable grants, which have covered travel expenses and fees for photography and the reproduction of illustrations in this book.

JAN BONDESON

London

The Feejee Mermaid

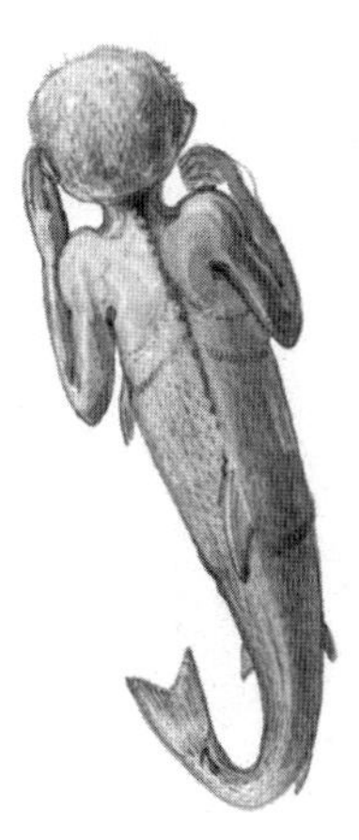

The Dancing Horse

If Banks had lived in olden times, he would have shamed all the enchanters in the world, for whosoever was most famous of them could never master or instruct any beast as he did.

Sir Walter Raleigh, *History of the World*

THE ANNALS OF PERFORMING ANIMALS STRETCH far back in time. Whereas some noble Romans kept small private zoos, the *profanum vulgum* had to content themselves with watching the dancing dogs and apes shown by itinerant jugglers. In medieval times, brutal animal baitings were a popular pastime, but the art of training animals became almost entirely forgotten. Bears, bulls, horses, wild boars, and badgers were baited with dogs; this degraded form of entertainment was relished by both high and low, and in Britain as well as on the European continent. Shakespeare mentioned the career of the fighting bear Sackerson, and the bears Tom o'Lincoln and Blind Robin were almost as famous. These brutal pastimes reigned supreme in medieval times and remained well into the sixteenth and seventeenth centuries. It was not until the late sixteenth century that the earliest animal trainer appeared for whom authentic records survive: the Englishman William Banks. His celebrated dancing horse Marocco was, in Elizabethan times, as well known throughout Britain and Europe, as any other two- or four-legged performer. Marocco even enjoyed the unparalleled honor, for a member of the equine race, of being mentioned in Sir Walter Raleigh's *History of the World*.

William Banks was born in Staffordshire, probably during the 1560s. One source describes him as a "Staffordshire gentilman," but it is more likely that he spent his youth as one of the retainers of the earl of Essex. In view of his great familiarity with horses, it would not have been surprising if his duties had included tending the earl's stables. His career as a horse trainer seems to have begun in the late 1580s. In the summer of 1591 he visited Shrewsbury with a performing white horse. He probably toured other cities, but apparently without gaining much renown. Some years earlier, probably in 1589, Mr. Banks had purchased a young bay foal, whom he had trained with the utmost care. The horse's name—Marocco—was derived from a type of saddle frequently used at the time. Marocco was a small, muscular horse with remarkable litheness and agility; he also proved particularly intelligent and easy to educate. Mr. Banks was much impressed with his horse's progress and had high hopes for his young charge. In spite of the novelty of their act and the consequent uncertainty of success—there are no previous records of any performing horse having been exhibited with profit—he decided to take up residence in London. Some time in 1592 or 1593, the adventurous Mr. Banks sold his belongings in Staffordshire, shod his horse with silver, and set out for the metropolis.

Exactly at what time the dancing horse first made his bow to a London audience has not been recorded; it is certain, however, that the shows were a great success from the start. Marocco could dance on either two or four legs with amazing agility. He could play dead in a particularly realistic manner. If Mr. Banks indicated some person in the crowd wearing a distinguishing garment, Marocco ran toward the person and pulled him into the arena. Sir Kenelm Digby wrote that Marocco "would restore a glove to the due owner after the master had whispered the man's name in his ear, and would tell the just number of pence in any silver coin newly showed him by his master." In an amusing trick, Mr. Banks ordered his horse to bow to the queen of England. Marocco did so with great reverence, ceremoniously scraping his hoof. But when he was ordered to bow to that archenemy of all Britons, the king of Spain, Marocco flatly refused. When Mr. Banks insisted, the horse neighed furiously, showing his teeth, and kicked out behind. The conclusion of this caper was that the infuriated horse, whose political opinions had been grossly insulted, chased its master out from the arena. This popular trick was aped by Marocco's successors among London's performing animals: a "jackanapes," an elephant, and Mr. Holden's old, unwieldy camel all shared Marocco's dislike of the Spanish monarch. The poet John Donne described, in one of his poems, a particularly apathetic and indifferent gentleman, who

> *. . . doth move no more*
> *Than the wise politique horse would heretofore*
> *Or thou, O elephant, or ape wilt do,*
> *When any names the King of Spain to you.*

Marocco was also famous for his arithmetic ability. Mr. Banks collected a number of coins from the audience and shook them in a large purse. Marocco then took up each of the coins in turn and returned them to their proper owners, after first stamping his hoof to tell how many shillings and pence each of them was worth. William Shakespeare mentioned the dancing horse's money-counting ability in *Love's Labour's Lost.* In the mid-1590s, William Banks and Marocco were among London's most popular entertainers. Mr. Banks was a wealthy man: he took lodgings at Belsavage Inn near Ludgate, where Marocco also had his stables. Their own arena was near Gracious Street, where Banks had ordered a gallery to be constructed. A musician was employed to entertain the spectators in between the shows and to play suitable music for the dancing horse. A merry tune, called Bankes' Game, was played to accompany Marocco's calisthenics. Some of the dancing horse's early tricks verged on the burlesque: in one trick Marocco drank a huge bucket of water and then relieved himself when ordered by the trainer. This amused one of Banks's literary friends, the poet John Bastard, who wrote that

> *Bankes has a horse of wondrous qualitie,*
> *For he can fight, and pisse, and daunce, and lie,*
> *And find your purse, and tell what coyne ye have:*
> *But Bankes, who taught your horse to tell a knave?*

The ladies of London were unamused by these unprepossessing antics, however, and Mr. Banks seems to have excluded this trick from his later repertoire. He also had Marocco's tail cut, making the horse a cut-tail or "curtall", as it was frequently described by his contemporaries. Another of Mr. Banks's literary friends, the famous author Thomas Nashe, wrote that "Wiser was our brother Bankes of these latter days, who made his jugling horse a Cut, for feare if at any time he should foyst, the stinke sticking in his thicke bushie taile might be noysome to his Auditors."

The ladies who attended Mr. Banks's later shows still had some unpleasant surprises in store for them. One of Marocco's foremost accomplishments was "to discern Maids from Maulkins." Mr. Banks ordered his horse first to fetch him a chaste and honorable virgin and then to bring him a harlot of the streets. It is unknown whether Mr. Banks trusted his own ability to determine the ladies' moral virtues from their clothing, or whether he let the horse select them at random. In either of these versions, this trick must have given rise to much coarse laughter from the male spectators. Another variation of this trick was described in one of the many anecdotes about the famous clown Richard Tarleton, about whom it was written that "for the part called the Clown's part, he never had his match, never will have." The authenticity of Tarleton's claim can be questioned, however, because Tarleton had been dead some years before Marocco came to London. Alternatively, the horse may have been older than earlier presumed, and identical to the "white horse" that Mr. Banks had brought to Shrewsbury in 1591.

The popular "Tarleton's Jests" related that once, when Tarleton and his actor friends were staging a play at the Crosse-Keyes near Gracious Street, Mr. Banks and his "Horse of Strange Qualities" were performing near the sign of the bell. Although the actors were wholly unsuccessful, Marocco had a field day, attracting much notice from the passersby. At last, Tarleton himself went up to see the dancing horse, and his friend Banks recognized him in the crowd. Mr. Banks asked the horse to seek out "the veriest foole" among the spectators. Marocco immediately ran to Tarleton and pulled him into the ring by the sleeve. Tarleton was laughed at by the crowd, but said nothing other than "God a mercy, horse!" He then said to Mr. Banks that he himself could make Marocco perform an even more startling feat, to which Banks replied that he would allow him to try, be it what it might. Tarleton told Marocco to seek out "the veriest whore-master" among those present: the horse ran to its own master and seized his coat with its mouth! As Marocco pulled Banks into the center of the arena, accompanied by the shouts and laughter of the audience, Tarleton said "Then God a mercy, horse, indeed!" It can be suspected, however, that the two cunning performers had planned the whole thing beforehand. Their advertising gimmick was a great success, and "God a mercy, horse" remained a byword in London for many years.

Late-sixteenth-century London was a hotbed of literary activity. Shakespeare's contemporaries were tireless in writing and publishing satirical pamphlets, squibs, and poems, and their work was eagerly bought by the growing, literate middle class. The satirical poems and pamphlets had strange titles—*Satiro-Mastix, Strappado for the Devill, Armin's Nest of Ninnies* and *The Mastive, or a Young Whelpe of the Olde Dogge*—many of which are difficult to comprehend today, owing to their extreme topicality. In November 1595, a laudatory poem dedicated to Marocco was printed. This "Ballad Shewing the Strange Qualities of a Yong Nagg Called Morocco" was apparently meant to be sold at the shows, but its contents are unknown, since not a single copy has been preserved to posterity. The month after, a thirteen-page pamphlet entitled *Maroccus Extaticus; or, Bankes Bay Horse in a Trance* was published in London. The alleged authors were Iohn Dando, the wier-drawer of Hadley, and Harrie Runt, head ostler of Bosomes Inne, but the pamphlet is likely to have been written by some Oxford undergraduates, who were among the main providers of contemporary satirical fiction. According to a nineteenth-century source, only two copies of this scarce pamphlet were known at the time. It was a common jest among the Londoners that Mr. Banks and Marocco probably could talk to each other, and the pamphlet has the form of a conversation between Banks and his horse. They talk about certain reprehensible features of daily London life and unanimously deplore the deceitful merchants and ungenerous publicans of the metropolis. The innkeepers were divided into two categories: those who were also keepers of brothels and those who employed prostitutes from the street. The pamphlet was illustrated with an amusing woodcut, showing Mr. Banks and his horse performing before some interested spectators.

After seeing Marocco in action, not a few spectators suspected that witchcraft was involved: no mere horse could perform such wonders. When Mr. Banks made an extended tour of the provinces from 1595 to 1597, visiting Oxford, Shrewsbury, and other cities, many people in the audience sat pale and trembling during the show; they were certain that Mr. Banks was a sorcerer and the horse his familiar spirit. In Edinburgh, Marocco again made a lasting impression, as judged from Patrick Henderson's *History of Scotland:* "There came an Englishman to Edinburgh with a chestain coloured naig, which he called Marocco. He made him do many rare and uncouth tricks, such as never horse was observed to do the like before in this land." The

Mr. Banks and his horse, a woodcut from the pamphlet *Maroccus Extaticus*, which was later reissued as a print. From the author's collection.

superstitious Scotsmen were certain that the horse was an evil spirit that would one day devour its master when his incantations could no longer harness it.

When Mr. Banks returned to London in the late 1590s, he found that his act had lost its novelty. London was full of performing horses, donkeys, apes, baboons, and bears. An elephant, imported to London, was a serious competitor, as was Mr. Holden's dancing camel, which had its arena on London Bridge itself. Extraordinary measures were needed to maintain Marocco's position as London's foremost four-footed entertainer, and the cunning horse trainer rose to the challenge. St. Paul's Cathedral was, at this time, London's centerpoint. The nobility strolled outside its gates, and the streets and shops nearby teemed with people. Jugglers, street pedlars, and beggars were everywhere. The cathedral did not have its present-day cupola, which was added after the Great Fire of London, but instead a very tall, square tower. From the summit of this tower, the entire city and its surrounding boroughs could be seen on a clear day; it cost a penny, paid to the verger, to be allowed to climb the tower. It is not recorded in the annals of St. Paul's, however, if there was a separate fee for animals: in February 1601, Mr. Banks led his horse up the more than a thousand steps of the endless spiral staircase out onto the church roof, which was, according to a contemporary writer, "as rotten as your great-grandfather." Here, more than 520 feet above ground, the horse danced and performed equilibristic tricks. Mr. Banks had of course spread the word about Marocco's great feat: all around the cathedral, people craned their heads not to lose sight of the horse up in the sky. Churchgoers, deacons, and clergymen rushed out not to miss this miracle. It was told, in a collection of anecdotes called *Jests to Make You Merie,* that a misanthropic old man sat in his chambers when his servant came running in, flustered and panting, to tell him about the horse standing "on the top of Powles" and the great multitudes of people being in the streets staring to behold it. The old man looked through his window to view the chattering crowd around the cathedral, replying "Away, thou foole, what need I goe so farre to see a horse on the top, when I can looke upon so many asses at the bottome?"

It was told, some years ago, in an American newspaper, that some drunken university students had brought a young bull up twelve flights of stairs into the flat of one of their friends—who, one might suppose, did not particularly appreciate this unexpected nocturnal visit. When the jokers tried to get the bull downstairs, it refused to move and instead became more and more furious. The neighbors called the police and the local television stations after their doors had been gored and kicked by the enraged animal, and the pranksters became national celebrities—probably welcoming their jail sentence, since it would enable them to evade the clutches of the animal rights activists. Some experienced policemen, who had previously encountered similar situations, were called. They decided that it was completely impossible to get any livestock to go down a flight of stairs, since that was against their nature. Instead, the bull was lassoed and pulled up another twenty-six flights of stairs, where the policemen planned to harness it to a helicopter and lift it off the roof of the building. The bull was frightened by the helicopter, however, and in spite of attempts to sedate it, the furious animal suddenly leapt from the roof. The bull nearly took one of the policemen, who held on to the lasso until the last second, with it on its headlong plunge toward the unyielding tarmac of the street below. Marocco's descent from St. Paul's cannot have been any easier. Defying the predictions of the twentieth-century Texan police officers, however, the horse, led on by his master, nimbly climbed down the endless stairway to receive the ovations of the masses below. Some years later, the poet Thomas Dekker considered this feat, in his *Guls Hornebooke:*

> From hence (the top of St. Paul's steeple) you may descend, to talke about the horse that went up; and strive, if you can, to know his keeper; take the day of the moneth, and the number of the steppes, and suffer yourselfe to beleeve verily that it was not a horse, but something else in the likenesse of one.

After the triumphant climb of the cathedral tower, the British Isles had become too small for Mr. Banks and his star performer, and he planned an extended tour of the continent. In March 1601, he set up headquarters at the Lion d'Argent Inn at Rue Saint Jacques in Paris. Under the new artist's name Monsieur Moraco, the horse made its debut some weeks later and was an immediate success: nothing even remotely like it had ever been seen by the Parisians. The amazement by which the feats of Marocco were received in the French capital has been graphically described by Monsieur Jean de Montlyard, Sieur de Melleray, the councellor of the prince of Condé. His eyewitness account of "cest incomparable cheval" was published as a long footnote to a French edition of *Les Metamorphoses ou L'Asne d'Or* of Apuleios; it describes Mr. Banks and Marocco at the summit of their extraordinary career.

The horse stood on two legs, walked forward and backward, and then knelt, extending his hooves straight out in front of him. He danced and capered with the agility of a monkey. Mr. Banks then threw up a glove, asking the horse to take it to a man wearing spectacles. The horse immediately did so. He then asked the horse to carry one glove to a lady wearing a green muff and another to a lady wearing a violet muff, to demonstrate that the horse knew colors; although there were more than two hundred people present, Marocco performed this task without a false step. When told to seek out a man with a bundle of papers under his arm, the horse did so although the man tried to hide the papers under his coat. Marocco used his strong teeth to seize the man by the cloak and pull him into the ring. Mr. Banks then blindfolded his horse and collected a large number of French coins in a purse. The horse, when asked how many coins were in the purse and how many of them were made of gold, gave correct answers by stamping his hoof. Mr. Banks then seized a golden écu from the purse and asked his horse how many francs such a coin was worth. Marocco stamped his hoof three times to mark that it was worth 3 francs, but did not seem quite satisfied with this answer. Prompted by Mr. Banks, Marocco then struck another four blows with his hoof, to designate that the écu was, owing to a recent change in the gold standard, now worth 3 francs and 4 sols. Chevalier de Montlyard was amazed by this: his imagination had to be stretched to the limit for him to believe that a horse could count money with its eyes blindfolded; then—"chose plus estrange"—it also knew the recent changes in the currency!

After this impressive demonstration of Marocco's power of intellect, some burlesque pranks were played to impress the more simple-minded spectators. Marocco neighed and sneezed when ordered to do so, showing his teeth and pricking up his ears. Whenever any spectator threw an object onto the stage, the horse brought it back like a spaniel. Mr. Banks then commanded his horse to walk as if carrying a lady, and Marocco ambled very gently around the arena. He then asked the horse to walk as if a riding master was mounted on him, and the horse leaped, scraped, bowed, and made the most intricate steps and passades. The jokes were continued when Mr. Banks harshly scolded his horse for being lazy, threatening to sell him to some carter who would soon work him to death, "et luy baillera plus de foüett de que fiun." Marocco hung his head and made other gestures to show his unhappiness. He then fell on the earth as if sick, rolling over with an agonizing groan. The horse lay absolutely still, playing dead with such skill that many spectators believed that he had really expired. Some rogues may have demanded their money back, while many soft-hearted people felt sorry for the poor horse, who had sacrificed his life for the sake of art. Mr. Banks then promised that the horse would revive if anyone would ask his pardon. Several spectators immediately cried out "Pardonnez-luy! Il fera bien son devoir!," and the horse jumped up, to everyone's relief. At Banks's beckoning, Marocco ran to a gentleman with red hair, who had been one of those interceding on his behalf, thanking his savior with many caresses. Mr. Banks then threatened to sell Marocco to the French postal service, who were not known to treat their horses kindly. Marocco raised up one leg and cantered about on three only, to show that he was lame and unfit for such arduous service.

Jean de Montlyard wrote that the show had once been visited by one of the city magistrates, who thought that such things could not be accomplished without magic. The horse and his master were both imprisoned for interrogation, but Mr. Banks managed to persuade their captors that the tricks were done only by art and signs, which the horse had been trained to obey since an early age. A distinguished visitor to the horse show was the philologist and philosopher Professor Isaac Casaubon. This famous scholar had visited several performances, becoming increasingly puzzled as to how this magic was achieved, if not by sorcery. Mr. Banks politely received Isaac Casaubon and managed to convince him that Marocco's feats were due only to his own careful training. He bragged that, given a year of preparations, he could train any other horse to perform similar tricks.

In the next year, Mr. Banks and Marocco arrived in Orléans, having probably visited several other French cities on the way. Their show was, once more, a great success. Orléans had several large Capuchin monasteries and churches, and the monks and priests were keen visitors to the dancing horse's performances. When treated to, one might suppose, a similar show to that performed in Paris, they were frightened out of their minds, calling out that Banks must be a sorcerer and the horse a demon from hell. Banks and Marocco were once more arrested and threatened to be burned alive, as witches or conjurers. To save himself and the horse, Mr. Banks demanded to be allowed to give a farewell performance before the priests and monks, which he was granted. Mr. Banks ordered Marocco to seek out one of the priests who had a large crucifix stuck in his hat. The horse did so, knelt down before the crucifix, and kissed it with the utmost piety. The monks and priests had to confess that they had made a mistake, since the devil did not have power to come near the cross. Instead, they said that the beast must have been inspired by the Holy Ghost, and they gave Mr. Banks "money and great commendations" when they left Orléans.

After this fortunate escape, Mr. Banks and Marocco continued their European tour for several years. It is likely that they performed in Lisbon, Rome, and Frankfurt. In the last city, Mr. Banks told an English cleric, Bishop Morton, about his adventures in Orléans. The Bishop later reproduced this tale in one of his theological pamphlets, *A Direct Answer to the Scandalous Exceptions of Theophilus Higgons,* written in 1609; it was later elaborated on in *The Booke of Bulls.* Mr. Banks's friend, the famous poet Ben Jonson, later added to this story by claiming, in an epigram published in 1616, that Banks and Marocco really had been burned at the stake:

But amongst these Tiberts, who do you think it was?
Old Bankes the juggler, our Pythagoras,
Grave tutor to the learned horse. Both which
Being, beyond sea, burned for one witch . . .

It is unknown exactly when Marocco retired from show business. Some writers have presumed that the horse might have died in some of the first years of the 1600s, but the horse was actually alive much later than that. According to certain recently discovered German documents, the (at least) 16-year-old Marocco was performing in Wolfenbüttel, Germany, in April 1605, at the court of Duke Henry Julius of Brunswick-Wolfenbüttel. This magnate, who was the brother-in-law of James I, was a considerable patron of British actors. In his account books, it is recorded that "Reichardt Banckes" received 40 thaler for the horse show, and the accompanying musician received 10 thaler. It is odd that this source, like Chambers' *Shakespeare,* gives Banks's first name as Richard; the evidence that his first name was really William rests solely on his vintner's license, signed William Banks. It is unknown whether Marocco died literally in harness, touring Europe until the last, or whether he was finally allowed to enjoy some years of retirement. On their return to England, like the great predecessor among their ilk, Marocco would have been able to view the manifold performing animals of London with a condescending horse laugh. It is likely that the dancing horse expired at some time in 1606 or 1607.

Mr. Banks, who had bragged that he could train another horse to take Marocco's place within a year, never again performed, in London or elsewhere. He had probably taken Marocco to court during their earlier residence in London, and it is recorded that in 1608 he received an appointment to the royal stables. He was paid a considerable sum from the privy purse of Prince Henry "for teaching of a little naig to vault, at his highnes comand." In the early 1620s, he was employed to train a horse for the duke of Buckingham. William Banks probably put away a good deal of money during his career as an international celebrity: in London, he was considered a wealthy and honest gentleman. It is unlikely that he had any shortage of interesting anecdotes from his travels. He had many friends, and his daughter married John Hyde of Urmstone, a Lancashire gentleman. William Banks was considered a great humorist by his contemporaries, but his recorded jokes have become rather dated. He once made a bet with the famous Moll Cutpurse, a notorious female pickpocket and dealer in stolen merchandise, that she would not dare to ride through London dressed in male apparel. This joke failed miserably, however, since she was attacked by a furious mob, full of moral indignation toward such an outrage against nature.

In 1632, when William Banks was about 70 years old, he decided to begin a new career. He wanted to open a tavern. This was a difficult venture in these days, since the Lord Mayor and Court of Aldermen of London had limited the number of licenses to keep "ale-houses," in a vain attempt to curb the prevalent drunkenness and hooliganism in the metropolis. Furthermore, Mr. Banks wanted to open his tavern in Cheapside, London's dominant mercantile street. After a year of bickering with the authorities, Mr. Banks went to the king and obtained a royal license to open his tavern. The year after, there was another quarrel, after the Lord Mayor had withdrawn his permission to serve food, but the king again supported the old horse trainer against the authorities. William Banks's tavern was among the most popular ale houses of London, and in 1637, some of his friends wrote a satirical squib about the strange foreign delicacies one might consume there. In *Shirley's Ball*, of 1639, a final mention is made of "mine hoste Bankes." This is the last record left by the old horse trainer, but it cannot be excluded that he was alive for several more years, to enjoy the success of his tavern. In 1662, Moll Cutpurse's biographer wrote that "I shall never forget my fellow Humourist Banks the Vintner in Cheapside who taught his Horse to dance, and shooed him with Silver."

When Marocco was touring Britain and France, many people were curious to know how the horse could be induced to perform such wonders, but Mr. Banks never divulged the secrets of his trade. In 1607, when he realized that he could never train another horse to become Marocco's worthy successor, he finally told a certain Mr. Gervase Markham exactly in what way Marocco had been trained. Markham was a country squire and an early hippologist; his book *Cavelarice* was set out to contain "all the Arte of Horsemanship, as much as is necessary for any man to understand, whether he be Horse-breeder, horse-hunter, horse-runner, horse-ambler, horse-farrier, horse-keeper, Coachman, Smith, or Sadler." Gervase Markham considered Mr. Banks's information to be of great interest, and "an explanation of the excellence of a horses understanding, and how to teach them to doe trickes like Bankes his Curtall" formed a separate chapter in the 1607 edition of *Cavelarice*.

From the time Marocco was a young foal, William Banks had spent most of each day together with his horse; to strengthen the bond between them, no other person was allowed to feed or caress Marocco. The horse followed Banks like a dog when he went about his daily business. William Banks always used kindness and patience during Marocco's lessons: when the horse performed well, it was rewarded with loaves of bread, but if Marocco showed obstinacy, he was given no food that day, as an incentive to become more attentive during next day's morning lesson. To teach Marocco how to count money, Banks first made the horse lift its leg on the command *Up!*, then indicated, by means of raising and lowering a rod in front of the horse, how many times it should stamp its hoof, "giving him a bit of bread til he be so perfit that, as you lift up your rod, so he will lift up his foot, and as you move your rod downeward, so he will move his foot to the ground." Soon, Marocco learned how to perform this trick without the use of a rod: as soon as the command *Up!* made him alert, Marocco could, by means of watching Banks's face, deduce how many times he should stamp his foot. William Banks's explanation of this behavior was that "it is a rule in the nature of horsses, that they have an especiall regard to the eye, face and countenaunce of their keepers." As soon as this trick had been perfected, it was easy for Banks to ask Marocco to tell him how many knaves, how many harlots, and how many rich men were in the audience.

To teach Marocco to bring a glove to any person in the audience, Banks first rewarded the horse when it brought back a glove that had been thrown. He then pointed his rod to an assistant and taught the horse to go to the assistant instead. Finally, several assistants were paraded in front of the horse, and the rod pointed toward one of them. If it looked as if Marocco would choose the wrong one, Banks called out *Be wise!*, and the horse at once chose a bystander instead. When the right person was selected, Banks called out *So, boy!* to indicate this. Finally, the horse became accomplished enough for Banks to direct him merely with his eyes. Marocco's ear for the different commands was soon sure enough for Banks to make them a part of his introduction for each trick, to prepare the horse for what was to be expected of him. Gervase Markham commented that Marocco's feats proved beyond doubt that a horse was as intelligent and teachable as a dog or jackanapes; it could perform the same tricks as these animals "except it be leaping upon your shoulders, climbing up houses, or untying knots, all which are contrary to the shape and strength of his great body." He had himself seen one of the shows, and he could well remember that the horse never removed his eyes from its master's face.

During the reign of Queen Elizabeth, the British Isles enjoyed a time of rapid economic development, with flourishing commerce and freedom from political tyranny and religious fanaticism. During the decades around 1600, there was also a rapid intellectual development, particularly in London. Large classes of society, recently liberated from poverty, had a gluttonous appetite for literature, drama, and other forms of entertainment. Seen from this perspective, Mr. Banks and Marocco were a phenomenon of their time, like William Shakespeare and other contemporary writers, artists, and dramatists. William Banks lacked any predecessor of importance: he was a great innovator within his particular field, and a man of genuine talent. It should be noted that Marocco's performances occurred in a time still redolent with witchcraft and barbarism. London had, at this time, a large "bear garden," where the most disgusting and cruel "sport" was advertised several times a week. Not only were bears baited, but also lions, bulls, and wild boar. A huge, three-story building contained more than 120 boxes for large mastiff dogs; new recruits to this company of fierce curs were needed, at regular intervals, since the baited animals frequently wrought havoc among their ranks. *The Elizabethan Stage* tells us that a horse was once chased around the arena by four hungry "mastives," with a screaming ape tied to its back. A visiting Spaniard found this dismal sight "most laughable." It is hoped that William Banks and Marocco were able to impress the Londoners with some of the respect for animal rights they were so obviously lacking. It was certainly a more edifying spectacle, from every point of view, to see the dancing horse perform than to watch an idiot devouring a living cat, a blind bear being whipped with a baboon tied to its back, or a badger with its tail nailed to the floor desperately trying to defend itself against four fierce fox terriers.

Marocco has had many successors. Perhaps the best known of these was the little horse Billy, one of the stars of Astley's circus in London during the late 1700s. Billy could dance, count, boil tea, and serve it like a waiter. After the circus had gone bankrupt, Billy was sold to a tradesman and had to pull a cart for several years until, one day, he was spotted by one of Astley's circus riders, who thought the dusty and rundown horse resembled the famous Billy he had once known. When the man clicked his fingernails as a sign for the horse to start tapping its foreleg, Billy did so at once. He was repurchased by the circus, which was once again solvent, and performed there for many years. After his death, at the venerable age of 42, Billy's skin was made into a huge thunderdrum that was used for special effects in the circus. In the late 1800s, the tricks of circus horses became increasingly dramatic. In 1885, the Italian Signor Corradini showed a horse walking the tightrope at the Theatre Royal in Covent Garden. His competitor Mr. Cottrell trained a mule to balance on a row of large bottles, which were, in their turn, balanced on a tightrope. Doc Carver, Buffalo Bill's former henchman, had a horse that could dive from a 36-foot platform into a huge basin full of water, whereas Freyer's Pony Circus boasted a team of horses that walked on stilts.

Marocco's intellectual achievements have also been challenged by his modern competitors. The best known of these was perhaps Clever Hans, a

Performing horses at the Circus Ambassadeur in Paris. One of the horses has the doubtful distinction of being ridden by a bear. From a poster in the author's collection.

robust German working horse that could answer questions and perform calculations by means of handling blocks adorned with letters or numerals. But although the German horse trainer Karl Krall believed that both Clever Hans and his predecessors, the Arab stallions Zarif and Muhammed, possessed a superior intelligence, the animal psychologist Dr. Oskar Pfungst dismissed his claims with scorn. It is likely that Krall directed his horses with hidden signs just as Banks had done, although this was never conclusively proved at the time. In the 1920s, the American parapsychologist Dr. J. B. Rhine was greatly impressed by Clever Hans's American counterpart, the

Professor Bristol's Equescurriculum: other late nineteenth-century rivals of Mr. Banks and Marocco. From Ricky Jay's private collection, reproduced by permission.

"Dummkopf! Drei mals zwei ist nicht fünf! Mach's noch einmal!" The Arab stallion Zarif is being taught by his stern German schoolmaster Karl Krall. Note the blinkers, used to focus the horse's attention on the blackboard. From an old photograph in the author's collection.

mare Lady Wonder; he alleged, in several scholarly publications, that the horse possessed supernatural gifts. Although these learned and acrobatic horses possessed considerable talents, none of them commanded even an inkling of the universal fame enjoyed by Marocco; it is doubtful if even the great modern circuses possess a horse capable of reproducing the tricks of its great sixteenth-century predecessor. Marocco will forever remain the only horse to have climbed the tower of St. Paul's Cathedral, since the steeple was reconstructed after the Great Fire of London: no hoofed animal will ever succeed in climbing the present-day cupola!

Mr. Banks and Marocco were mentioned more than sixty times in the contemporary literature: already during its lifetime, this miraculous horse had become a character of legend. The most exaggerated tales were current about Marocco's acrobatic tricks "on top of Powles." Through Ben Jonson's garbled account, several pamphleteers and scandal chroniclers repeated the untrue tale that Marocco and his trainer had been burned at the stake by the papists. In a French humorous pamphlet, published in 1626, the English houyhnhnm is one of the main characters: "Le joly Monsieur Maroc" plays the part of Dante's Virgil, residing in the nether regions of hell after the conflagration, and entertaining a visiting Frenchman with great eloquence. Another doubtful authority, the apocryphal mock romance of Don Zara del Fago, goes into further details about these matters:

> Banks his beast; if it be lawful to call him a beast, whose perfections were so incomparably rare, that he was worthily term'd the four-legg'd wonder of the world for dancing; some say singing, and discerning maids from maulkins, finally, having of a long time proved himself the ornament of the British clime, travailing to Rome with his master, they were both burned by the commandment of the Pope.

In 1654, when both Banks and Marocco had been dead for several years, the poet Edmund Gayton wrote his *Pleasant Notes to Don Quixot,* a versified introduction to this popular novel, which was quite a best-seller in mid-seventeenth-century Britain. In one of the poems, Marocco makes a long speech to Don Quixote's horse Rosinante, extolling his own virtues, which went far beyond those of the Spanish knight's faithful steed. The famous dancing horse with silver shoes, who could count money by stamping its hoof and who climbed the tower of St. Paul's Cathedral, while Rosinante was only hoisted up a windmill, provides a suitable conclusion to this tale:

> *Though Rosinante famous was in fields*
> *For swiftnesse, yet no horse like me had heels.*
> *Goldsmiths did shoe me, not the Ferri-Fabers;*
> *One nail of mine was worth their whole weeks labours.*
> *Let us compare our feats; thou top of nowles*
> *Of hills hast oft been seen, I top of Paules.*
> *To Smythfields horses I stood there the wonder;*
> *I only was at top; more have been under.*
> *Thou like a Spanish jennet, got in the wind,*
> *Wert hoisted by a windmill; 'twas in kinde.*
> *But never yet was seen in Spaine or France,*
> *A horse like Bancks his, that to the pipe would dance:*
> *Tell mony with his feet; a thing which you,*
> *Good* Rosinante *nor* Quixot *ever could doe.*

Lament of the Learned Pig

O Heavy day! Oh day of woe!
To misery a poster,
Why was I ever farrowed—why
Not spitted for a roaster?

Of what avail that I could spell
And read, just like my betters,
If I must come to this at last,
To litters, not to letters?

THE AUTHOR OF THIS MISERABLE LAMENTATION is Toby, the Learned Pig. This once famous performing pig is the hero of one of Thomas Hood's many humorous poems, which was originally published in 1835, in his *Comic Annual.* In his time, Thomas Hood was a poet of considerable repute. He wrote witty and apposite comments to current scandals and political affairs, and his punning rhymes and clever metrics were highly regarded by his contemporaries. In his *Collected Works,* the poem about the learned pig is accompanied by such popular favorites as *Miss Kilmansegg and Her Precious Leg* and *Faithless Nelly Gray.* Although some critics blamed Thomas Hood for his frivolousness and shallow nature, his poetry remained popular well into the early twentieth century; it has become one of the casualties of the present age, containing neither fashionable psychological "truths" nor incisive social criticism. *Lament of Toby, the Learned Pig* is about a famous performing pig which, at the end of its distinguished career, suffers the dire fate of being fattened for the table after no longer being able to earn its keep in the daily shows. It may well be based on fact, since these learned pigs were no rarity in Thomas Hood's time.

In this world, pigs, as well as men,
Must dance to fortune's fiddlings,
But must I give the classics up,
For barley-meal and middlings?

At least since the sixteenth century, performing animals were exhibited at the old English fairs and markets. The miraculous dancing horse Marocco had many seventeenth- and eighteenth-century successors, but none of these made history in the same way. In the second part of the 1700s, there was a rising interest in performing animals, and the London spectators had much to choose from: vaulting apes, dancing dogs, counting horses, drumming hares, and Chinese starlings playing cards. There was a great variety of flea circuses, whose insects performed the most astounding wonders. A British troupe of fleas enacted the siege of Antwerp with cannon the size of pins, and portrayed an elegant ballroom scene with several dancers and an orchestra of twelve. Herr Lidusdroph's company of Learned and Industrious Russian Fleas boasted a star performer—the Russian Hercules, who could carry twelve of his fellows on his back. For the real connoisseur, there was the quack Colonel Katterfelto's regiment of electric black cats and Mr. Breslaw's famous acting canaries. In one of the latter's theatricals, these remarkable birds, all dressed in military uniforms, shot an alleged deserter with a small cannon and then buried the corpse, whistling mournful funeral tunes. Those who sought further amusements after viewing this parliament of wonders would be delighted to find out that, to quote from a contemporary advertisement, "The Wonderful Dog will play any Gentleman at Dominoes that will play with him."

O, why are pigs made scholars of?
It baffles my discerning,
What griskins, fry, and chitterlings
Can have to do with learning.

The first man ever to think of training pigs for the stage was the shoemaker Samuel Bisset, who was born in Perth, Scotland, in 1721. After marrying a wealthy woman, he left his shoemaking shop for good and later moved to London to set up business as a broker. One day, when Mr. Bisset was more than 50 years old, he read about a "thinking horse" that had performed at St. Germains, and decided to change his career once more. He purchased a horse, a dog, and two monkeys and became a full-time animal trainer. Some months later, the animals were ready for their first appearance in public. Apparently, they did quite well, in spite of their somewhat elderly trainer's lack of zoological expertise and the stiff competition from the host of other four-legged artists in London. Later, Mr. Bisset toured the provinces with a troupe that also contained two monkeys walking the tightrope and playing the barrel organ, a hare beating the tabor, canaries that could spell, and a tortoise that could bring back objects like a dog (although more slowly!). The star performers were his Cat's Opera: an orchestra of cats sitting with music books in front of them, beating on dulcimers and drums with their paws, "squalling at the same time in different keys or notes". When these cats performed in London, at the Haymarket, during the next season, they attracted much attention. In little more than a week, the Cat's Opera brought in nearly 1,000 pounds to their owner, a sum illustrating the Londoners' great delight in animal amusements.

Mr. Bisset was continuously widening his repertoire and developing new tricks. After having trained many different species, he asked himself "whether the obstinacy of the pig might be conquered." Early in 1782, he purchased a young black piglet at the market in Dublin, paying 3 shillings for the animal. After 6 or 7 months, it had made but little progress. Lesser men would have despaired, but Mr. Bisset went on for another 16 months before the pig was sufficiently accomplished to make its first appearance, which was probably in Dame Street, Dublin. Although the mentor of the learned pig sometimes stooped to using cruel tricks—the Dancing Turkeys were only waltzing when a section of the stage was heated under them!—he seems to have used humane methods during his porcine pupil's long education. The poet Robert Southey, who was interested in learned pigs, once interviewed a man living near the pig trainer's backyard: "He told me that he never saw the keeper beat him; but that, if he did not perform his lessons well, he used to threaten to take off his red waistcoat,—for the pig was proud of his dress."

My Hebrew will all retrograde,
Now I'm put up to fatten,
My Greek, it will all go to grease
The dogs will have my Latin!

The Wonderful Pig soon became the leading light of Mr. Bisset's troupe. It could kneel, bow, spell out names using cardboard letters, cast up accounts, and point out married and unmarried people in the audience. It was a great novelty among the Dubliners, and the shows were always well attended. Mr. Bisset earned a considerable sum of money. His heart must have swelled when he thought of his pig's great expectations: after a triumphal tour of the Midlands, it would enter London like a conqueror. But fate had other plans for the old animal trainer. The Dubliners demanded further performances, and when Mr. Bisset moved to another district of the city, he forgot to ask the permission of the local officials, an oversight that would cost him dearly. During the pig's performance, a policeman sent by the local magistrates entered to disrupt the show. When old Mr. Bisset resisted, he was given a savage beating, and all the stage equipment was destroyed. The brutal con-

The original Learned Pig performing in London, as drawn by Thomas Rowlandson in 1785. From a print in the author's collection.

stable swore that if the Learned Pig was ever seen again within his jurisdiction, he would slaughter it on stage and drag its owner to prison. The elderly pig trainer took to his bed, for several days, after this unprovoked assault. Some weeks later, he left Dublin, taking his pig and other animals with him. He planned to tour England en route to London, but in Chester, Mr. Bisset fell ill and died. According to his biographers, the meek old pig trainer never quite recovered from the brutal assault in Dublin.

❧

Alas! my learning once drew cash,
But public fame's unstable,
So I must turn a pig again,
And fatten for the table.

Mr. Bisset's learned pig was taken over by a certain Mr. Nicholson, about whose previous career little is known. It is likely that he was some kind of impresario or theatrical manager, since he seems to have known his way about show business. At any rate, he knew well that he had come on to a good thing: under his management, "the grunting Philosopher" was certainly not kept standing idle. They first went on tour to Scarborough, York, and several other cities. As Nicholson's troupe also boasted a hare that beat the drum, a tortoise that fetched and carried like a dog, and six turkey cocks that performed a country dance, it is likely that he had purchased Mr. Bisset's entire menagerie. On February 16, 1785, the London newspapers announced that the porcine prophet was among the celebrities recently arrived in town. Nine days later, Mr. Nicholson secured a suitable apartment for his pig, at Charing Cross opposite the Admiralty. In an advertisement, it was pointed out that "this sagacious animal reads, writes, and casts accounts, by means of typographical cards; solves questions in the four rules of arithmetic, tells, by looking at any gentleman's watch, what is the hour and minute, &c, and is the admiration of all who have seen him."

The Learned Pig had been a great novelty in Dublin, but it caused even more astonishment in London, soon ruling its contemporary four-legged performers like a sovereign. Every day, huge crowds came to see the pig. Mr. Nicholson was no philanthropist, and he made the porcine prophet work hard. The shows were frequent, and some newspaper men felt pity for the poor pig, particularly since the April weather was uncommonly hot: "what with the weather, and the concourse of visitors, the poor animal is so *roasted,* that its skin is almost a *crackling.*" All classes of society were united in their admiration for the grunting Philosopher, and all religious persuasions. According to a rather sneering notice in the *Morning Post* of April 26, even "some of the *Synagogue Beaux* made a party to see the learned Pig. Tho' they were, on the whole, entertained with his grunting honour, their ideas of his merits were rather lessened, as the *good peoplesh* have always believed that the Devil inherited a herd of Swine."

In April, Mr. Nicholson was prevailed on to proceed to Oxford, Bath, and Bristol, but he publicly declined this offer, since the Learned Pig's reign in London lasted longer than even he had calculated. In July, the pig joined the circus performing at Sadler's Wells Theatre. A review in the *Morning Herald* stated that "he went about his business in a capital style, to the admiration of a crowded house, and was honored with the loudest plaudits from every part of it." The pig's fame angered the troupe of tightrope performers and acrobats, and the theater manager, Mr. Richard Wroughton, soon received a deputation consisting of Signor Placide, Mr. Redigé (otherwise known as the Little Devil), Monsieur Dupuis, Mr. Meunier, and La Belle Espagnole, all thinking themselves degraded by being forced to perform together with a pig. When the manager tried to persuade them to stay, the furious acrobats asked him to choose between the pig and them. They were all sacked, on the spot, and the pig had the arena to itself. The Learned Pig's reign of London's amusement world continued until the early autumn, when it at last departed for Oxford and Bath, the first stops of its grand tour of Britain.

Day after day my lessons fade,
My intellect gets muddy;
A trough I have and not a desk,
A stye—and not a study!

In December 1785, the Learned Pig performed at Norwich, and one of the shows was seen by the Reverend James Woodforde, who described it in his *Diary of a Country Parson:*

> After Dinner the Captain and myself, went and saw the learned Pigg at the rampant Horse in St. Stephens—there was but a small Company there but soon got larger—We stayed there about an Hour—It was wonderful to see the sagacity of the

> Animal—It was a Boar Pigg, very thin, quite black with a magic Collar on his Neck. He would spell any word or Number from the Letters and Figures that were placed before him; paid for seeing the Pigg 0.1.0.

Parson Woodforde was not the only observer to note that the "Pigg" was particularly thin; it is likely that the clever Mr. Nicholson purposely kept it lean to increase its agility during the shows.

In March 1786, the pig was back in London for a second sojourn in the metropolis, performing at the Lyceum, in the Strand. His show was unchanged, except that it was pointed out that "he never divulges the thoughts of any lady in the company *but by her permission*"; this change might have been due to an embarrassing incident during a previous thought-reading session! Again, the pig was a great success, and the poet Robert Southey wrote that the Learned Pig was "a far greater object of admiration to the English nation than ever was Sir Isaac Newton." In May 1786, the grunting Philosopher left London for another lengthy tour, and visited Banbury and Oxford before going to France in the winter of 1786. A newspaper account from January 1787 tells us that "the Learned Pig of Charing-cross is one of the rarer monsters of France. It has fed its owners fat through Calais, Boulogne, Montreuil, &c., &c." In the meantime, several other learned pigs appeared in London, to reap some of the fame of their departed colleague. "The Amazing Pig of Knowledge" appeared at the Crown Inn, and Astley's circus had a French pig, which "articulated *oui, oui* with an uncommon fine accent," while Hughes's circus boasted an automaton pig, imported from Italy. The original porcine prophet went for another extended tour of the English countryside in 1787, visiting Retford, Newark, Lincoln, and Stamford, among many other towns. The Learned Pig surfaced in Edinburgh in late 1787, where it was seen by Robert Burns. When Burns was asked to attend a party for the first edition of his poems, he answered that he would, on condition they also had the Learned Pig present; he apparently did not want to put himself on a show to advertise his work.

During its long and distinguished career, Mr. Nicholson's learned pig was stated by the newspapers to have earned more money "than any actor or actress within the same compass of time." In November 1788, articles in several newspapers informed the public that their old friend and favorite, the Learned Pig, had departed this life. The same article also mentioned that its master was confined in a madhouse at Edinburgh: "Too much learning we

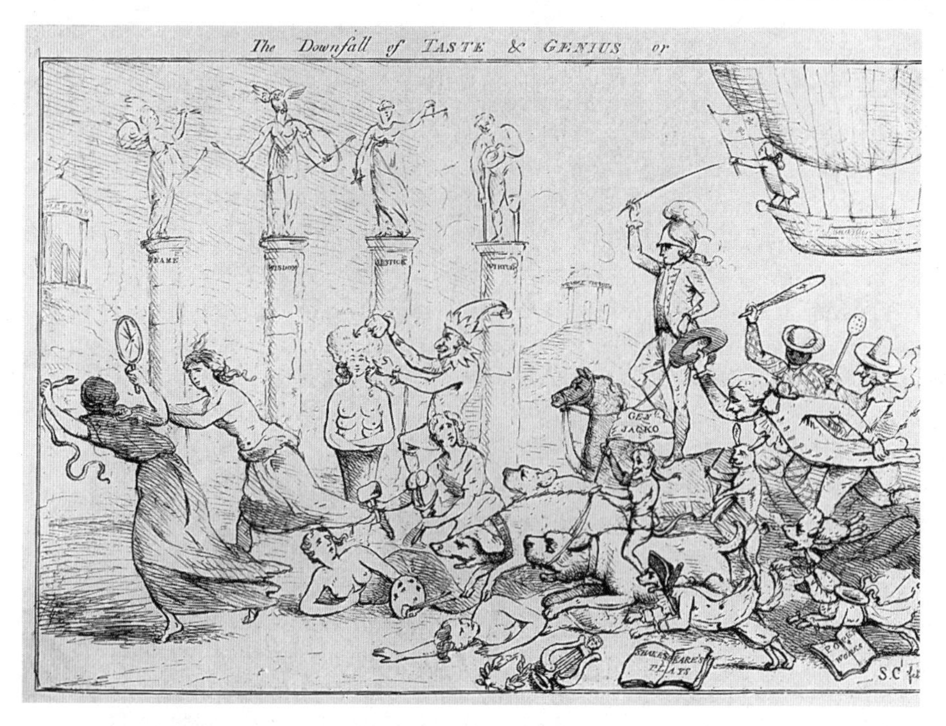

The caricature *Downfall of Taste and Genius*, or *The World as It Goes*, by Samuel Collings, depicting the muses of art and music being assaulted and trodden underfoot by a host of trained animals, clowns, and burlesque comedians; the Learned Pig is leading the charge. ©The British Museum. Reproduced by kind permission.

suppose had driven the pig mad, so he bit his master!" This leaves unexplained some newspaper articles from October and November 1789 (included in Lysons' *Collectanea*) stating that a learned pig, allegedly the same one that 4 years since "afforded such amusement in most parts of England," had recently returned to Hereford after a long and arduous tour of France. The grunting Philosopher had witnessed the French Revolution at first hand, and a advertisement bragged that "from his frequent interviews with the French patriots, he is almost enabled to hold a discourse upon the Feudal System, the Rights of Kings, and the Destruction of the Bastile." The last heard of this revolutionary learned pig—whether a pretender or Mr. Nicholson's original animal under new management—is that it departed for Monmouth and Abergavenny after its 10-day stay in Hereford.

Of all my literary kin
A farewell must be taken
Good Bye to the poetic Hogg
The philosophic Bacon!

Although neither Sir Francis Bacon nor the Scottish poet James Hogg would have been disposed to consider a pig his equal, Toby's talk of his "literary kin" is not without foundation, since many late-eighteenth-century writers mention learned pigs in their work. According to Boswell, Samuel Johnson once discussed learned pigs. In conversation, his friend Miss Seward remarked that she had seen a performing pig in Nottingham in 1784; this was probably Mr. Nicholson's animal on its way to London. The subject amused Dr. Johnson, and he declared that "pigs are a race unjustly calumniated. Pig has, it seems, not been wanting to man, but man to pig. We do not allow time for his education, we kill him at a year old." One of those present feared that brutal methods must have been used to subdue the inherently indocile animal, but Samuel Johnson thought that the pig had no reason to complain, since it had escaped the slaughter: "protracted existence is a good recompense for a very considerable degree of torture." After Dr. Johnson's death, an old enemy wrote an ironic epitaph:

Though Johnson, learned Bear, is gone,
Let us no longer mourn our loss,
For lo, a learned Hog is come,
And Wisdom grunts at Charing Cross.

Happy for Johnson — that he died
Before this wonder came to town,
Else it had blasted all his pride
Another brute should gain renoun.

The poet William Cowper was compared with the Learned Pig by a waggish correspondent, whereas another likened his fame to that of the actress and courtesan Anne Bellamy. Cowper responded that "You tell be that I am rivalled by Mrs. Bellamy and he that I have a competitor for fame not less formidable in the learned Pig. Alas! what is an author's popularity worth, in a world that can suffer a prostitute on one side, and a pig on the other to eclipse

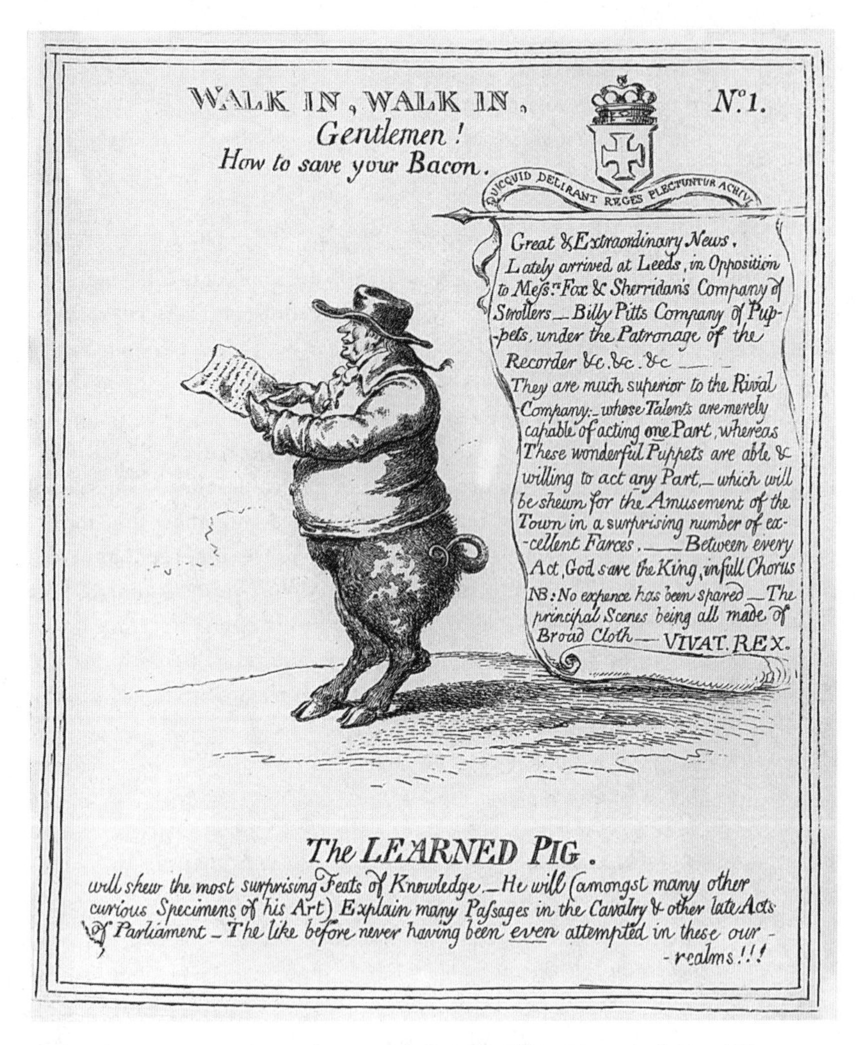

One of the many caricatures mocking Prime Minister William Pitt as the Learned Pig. ©The British Museum. Reproduced by kind permission.

his brightest glories." In the humorous journals, Prime Minister William Pitt was often referred to as the Wonderful Pig, and some wags even agitated that the pig should run for Parliament. Several satirical drawings depicted Pitt as a pig, with or without learned attributes. One drawing, entitled *The Rival Pigs,* showed William Pitt and his main political adversary, Charles James Fox, with other learned pigs in the background, some of them wearing wigs.

O why did I at Brazen-Nose
Rout up the roots of knowledge?
A butcher that can't read will kill
A pig that's been to college.

Some strictly religious men were deeply affected by the Learned Pig's performances and suspected black magic to be involved. They even demanded that the pig should be burned and its manager banished, since they did not doubt that he was "corresponding with the Devil." Other contemporary thinkers considered the pig as proof of the transmigration of souls, suspecting that "the Spirit of the grunting Philosopher might once have animated a Man." Several contemporary scholars took a more constructive interest in learned pigs and other performing animals, using them as arguments in the age-old philosophical question of whether animals possess reason. The extreme Cartesian view that all creatures great and small were nothing more than automata, whose actions were merely directed by instincts, was challenged by many other seventeenth-century philosophers. When the Cambridge scholars Matthew Wren and John Preston disputed, in front of King James I, the question of whether dogs could reason, it was one of the earliest attempts to solve this old philosophical conundrum. Although John Locke did not consider beasts to be capable of comparing ideas or reasoning abstractly, David Hartley thought that a clever and well-trained dog might well be compared to a deaf-mute or a retarded human being, who could communicate only by signs. David Hume also had a surprisingly high opinion of the brute creation's intellectual abilities, which might well have been influenced by the many performing animals of his time. He made no sharp distinction between human reason and animal instincts: the difference might instead be that the human being had a larger number of inherent instincts, one of which was the capacity for deductive reasoning.

The eccentric Lord Monboddo, an early evolutionist who was much mocked by his contemporaries, had similar ideas, which went a bit too far: he believed that the orangutan and the human being belonged to the same species, and that humankind's greater intellectual ability was merely a result of their higher social development. Like Jean Jacques Rousseau, he thought that the language was an invention of society rather than an innate human attribute separating them from animals. Lord Monboddo had spoken to several experienced animal trainers in London, and they all agreed that the brute creation might well approach the human beings with regard to intellect, if they lived long enough and if sufficient pains were taken regarding their education. Charles Darwin's grandfather, Dr. Erasmus Darwin, was also influenced by the Learned Pig. In his *Zoonomia,* published in 1796, he wrote that the pigs were a much-maligned race: had they been allowed to lead an interesting and variable life instead of being locked up in their styes, and left to develop intellectually instead of being slaughtered at an early age, their position within the animal world would have been an exalted one.

In the popular eighteenth-century natural history books, the animals were often anthropomorphized. The lion was noble and lordly, the dog loyal and servile, the cat wanton and deceitful. For many years, the pigs were reviled in these natural history books: in spite of their considerable value as food animals, they were castigated as being gluttonous, selfish, and sordid. That a filthy pig, a slovenly, dull and "pig-headed" brute could not only perform tricks that had previously been mastered by apes and dogs but also read thoughts and play cards was a great astonishment to many people. This would explain much of the Learned Pig's fame before its contemporaries.

One thing I ask—when I am dead,
And past the Stygian ditches—
And that is, let my schoolmaster
Have one of my two flitches.

'Twas he who taught my letters so
I ne'er mistook or missed 'em
Simply by ringing *at the nose,*
According to Bell's *system.*

The pig trainer Samuel Bisset was, like William Banks, the great forerunner in his particular line of business. Indeed, there had been trained pigs before his time, but their success had been scant. Most people disliked these foul, smelly brutes, and their tricks had not been impressive. The eccentric King Louis XI of France was an exception: he preferred pigs to all other animals and always kept many pet pigs at his court. Once, as a joke, he ordered the learned abbot of Baigne to teach these pigs to sing in concert. The abbot took the challenge seriously, and retired to his country seat to focus on the task. None of the courtiers believed that he would make it, but after a couple of months of inventive thinking and hard labor, the abbot returned with a large and complicated apparatus. All the pigs were seated on stools, in strict order of age and heaviness. The court sat down in a comfortable gallery in front of the pig orchestra. When the king had been seated on his throne, the abbot started playing the remarkable organ he had brought with him: each time a key was touched, a sharp spike, arising from the stool's mechanism, pricked the behind of the pig sitting on it, and the pig yelped. The squeals of the young piglets were more shrill than those of the old, heavy sows, and the abbot was able to play pieces of music to the king's entire satisfaction.

It is unknown exactly what methods Samuel Bisset used when training his Learned Pig, but it is unlikely that the Scottish doctor Joseph Bell's educational system, mentioned by Toby in his long lamentation, was in any way involved. Bisset had no pupils, and his untimely death rendered it impossible for him to impart his secret training methods to anyone. From 1797 to 1798, another learned pig was exhibited in the United States by the British conjurer W. F. Pinchbeck, who later sold the pig at a public auction. In his book *The Expositor, or Many Mysteries Unravelled,* Pinchbeck described how he had trained a young piglet to move cards with letters on them: he scolded it in a loud voice when it had made a mistake and likewise rewarded it with pieces of apple when it had done its work well. To get the animal to move the card into its correct position to form words, he led it on a leash, teaching it to obey certain hand signs, until he could direct it using only these signs. The same technique, albeit with figures instead of letters on the cards, was used to teach the pig to count and read a watch. Pinchbeck considered pigs to be both clever and ready to learn: "the animal is so sagacious that he will appear to read your thoughts." Pigs have excellent hearing, which Pinchbeck used to his advantage, by means of directing the animal with snuffling, clearing his throat, and snapping his fingers.

Even today, trained pigs appear at various circuses, although their feats are acrobatic more often than scholarly. The American pig trainer Bobby Nelson shares Drs. Johnson and Erasmus Darwin's opinion that the pig is an underestimated animal. Its intelligence and excellent eyesight, hearing, balance, and sense of smell make it easy to train if it is taken early from its mother. The sole problem is that the pig's propensity to fatten tends to make it too clumsy for acrobatics by the time it is only a few years old. A notable late-nineteenth-century pig trainer was the Russian clown Vladimir Durow, who usually made his entrance seated in a cart drawn by his huge sow Tschuschka. In a spectacular trick, the pig stood on top of a pedestal, a male goat stood on Tschuschka's back, a greyhound on that of the goat, and a cat on the dog's back. This pyramid of animals was crowned by a large rat, sitting on the cat's back. The pig's virtues in this trick were its solidity and the strength of its legs rather than any equilibristic ability. The acrobatic English piglet Bill, belonging to Lord George Sanger's circus, was Tschuschka's absolute opposite: he could balance on one trotter, standing on the nose of a sea lion. In the 1950s, the Danish pig trainer Carl Hansen had considerable success with his pig Hansi, which was capable of playing the glockenspiel and balancing on huge barrels (although not at the same time!). The Swiss clown Michels Ghazzi was much less fortunate as a pig trainer. He had taught a pig to hold a pistol in its muzzle and fire it at a target, but the pig turned unexpectedly and shot the clown instead; the unhappy Swiss expired a few days later, owing to blood poisoning.

❧

For sorrow I could stick myself,
But conscience is a clasher;
A thing that would be rash in man
In me would be a rasher!

It took until 1817 before a learned pig approaching the fame of Bisset and Nicholson's original porcine performer stood on a London stage: it was Toby, the Sapient Pig, who was performing at the Royal Promenade Rooms as a part of the show of the conjurer Mr. Hoare. Toby was always dressed in a large lace collar during the shows—attire that apparently belonged to the dress code among learned pigs—and collars of this kind were known as "Toby frills." Like his predecessors, Toby could spell, read, and count; he could also read thoughts and tell the spectators' ages. His foremost showpieces were to play cards, read a watch, and guess the number of objects in a closed box held before him. The pig was soon popular enough to have the conjuring show all to himself, and to promote the pig's antics further, the clever Mr. Hoare sold Toby's autobiography for 1 shilling. This slender volume was adorned with the portrait of its alleged writer, reposing in his literary pigsty, as a frontispiece. It told the story of its author's life from his first step until his present exalted position as an entertainer. Toby wrote

The frontispiece of Toby's autobiography. From the author's collection.

that his expectant mother had strayed into a well-stocked library and devoured many of the books. The bookish sow had thus, in accordance with the contemporary embryological ideas, laid the foundations for his precocious young mind, always eager for more learning. The pig also described his youth and schooling with Mr. Hoare: his name had of course been derived from Hamlet's soliloquy—*to be* or not *to be!* Toby's early experiences of show business had been disappointing, since drunken yokels had heckled and jeered the porcine scholar, but he eulogized the educated London public who had became his patrons.

Toby's greatest competitor was another learned pig, which was alleged to be equally well read, having studied with a famous Chinese philosopher and then toured Latin America through the good offices of the Spanish Inquisition. The manager of this widely traveled pig, who apparently did not share his charge's penchant for correct spelling, promised 1,000 guineas to anyone who could show the like of "this Pillar of Pythagorus [*sic*], and the wonder of the present age." Competition was keen between the rival pigs, but the outcome was that the usurper lost both his audience and his life: the proprietor had his artist slaughtered, despairing of the unwieldy hog's future

A good drawing of Toby performing, from the Finninham Collection of Cuttings from Newspapers. Reproduced by permission of the British Library.

prospects as an entertainer. In contrast, Toby arrived at a ripe old age. According to advertisements kept in the British Library, he was still going strong in 1823, and probably in 1825.

To leave my literary line
My eyes get red and leaky;
But Giblett doesn't want be blue
But red and white, and streaky.

The ultimate fate of the "Unrivalled Chinese Swinish Philosopher, Toby the real Learned Pig," London's last learned pig of importance and the ultimate in a distinguished line of performers, was much less edifying. This was the pig that inspired Thomas Hood's poem. From 1833 until 1835, it was exhibited in a small-time London animal show owned by a man named Mullins. Thomas Hood was a frequent visitor to circuses, fairs, and markets, and the sights often inspired his poetic imagination. He wrote his *Ode to the Cameleopard* after seeing a giraffe exhibited, and another of his poems is dedicated to the famous clown Joseph Grimaldi. None of the available biographical works on Thomas Hood mentions that he visited the pig, but it is still likely that he did so, and that he was later informed about the pig's sad fate through the newspapers. Few records have been kept about poor Toby, and it is likely that his fame and skill did not compare favorably with those of his two great predecessors. While employed by a circus in 1833, Toby acted together with the manager's "monstrously fat child," perhaps indicating that he was a less attractive performer than the earlier learned pigs. Toby was finally sold to the butcher Mr. Giblett, under the dismal circumstances detailed by Thomas Hood:

Old Mullins used to cultivate
My learning like a gard'ner;
But Giblett only thinks of lard,
And not of Doctor Lardner.

He does not care about my brain
The value of two coppers,
All that he thinks about my head
Is, how I'm off for choppers.

The Feejee Mermaid

Mermaids yet have never been seen.
One there was in Piccadilly,
Half a fish, and half an ape;
You must think me very silly
To believe in such a shape.

Anonymous, *The Comic Almanack* II (London 1844)

IT BELONGS TO HUMAN NATURE TO STRIVE TOWARD the impossible: to fly in the air with the birds, or to swim underwater like the fishes. The belief in a subspecies of fish-tailed human beings is age old: many ancient legends mention water-living—predominantly female—semihuman creatures with a human face and a fish's tail. As a mythological figure, the mermaid has survived well into our time, although the fantasies of Hans Christian Andersen and Walt Disney & Co., and the seductive mermaid on the cover of a can of tuna fish may seem far removed from the fish-tailed Phoenician moon goddess Atergatis. For several hundred years, the mermaid has been praised by poets and bards, being described as a beautiful woman down to the waist but with a long, scaly fish tail. She was wont to admire her appearance in a mirror while combing her long hair, as she does in Lord Tennyson's *Confessions of a Mermaid:*

I would comb my hair till my ringlets would fall
Low adown, low adown,
From under my starry sea-bud crown
Low adown, and around.
And I would look like a fountain of gold
Springing alone,
With a shrill inner sound,
Over the throne
In the midst of the hall.

The song of the mermaid was believed to have a hypnotic power over both man and beast. In Shakespeare's *A Midsummer Night's Dream,* Oberon tells Puck that

Once I sat upon a promontory,
And heard a mermaid on a dolphin's back
Uttering such dulcet and harmonious breath,
That the rude sea grew civil at her song,
And certain stars shot madly from their spheres
To hear the sea-maid's music.

Unlike his distinguished poetic predecessors, Charles Dickens did not think highly of mermaids. To the confusion of readers and literary historians alike, the references to mermaids in his novels are often derogatory, not to say insulting. In *The Old Curiosity Shop,* Dick Swiveller discusses the character of his employer's ungenerous and bad-tempered wife with the words, "I wonder if she *is* a dragon by-the-bye, or something in the mermaid way." In *Barnaby Rudge,* it is coldly observed that "according to the constitution of mermaids, so much of a mermaid as is not a woman, must be a fish." An even more mortifying insult to the mermaid tribe occurs in *Nicholas Nickleby,* in Chapter 42, when Miss Fanny Squeers, the daughter of the notorious Wackford, the schoolmaster of Dotheboys Hall, has a violent quarrel with her friend 'Tilda: " 'This is the hend, is it?' continued Miss Squeers, who, being excited, aspirated her h's strongly; 'This is the hend, is it, of all my forbearance and friendship for that double-faced thing—that viper, that—that *mermaid?*' (Miss Squeers hesitated a long time for this last epithet, and brought it out triumphantly at last, as if it quite clenched the business.)" Many Dickensians may have wondered why the great novelist had such a low opinion of mermaids, but the explanation is likely to be, surprisingly enough, that he had seen one himself. In 1822, a dried mermaid was exhibited at the Turf coffeehouse in London. It made a great sensation in town, and although the price of admission was as high as a shilling, the 10-year-old Charles Dickens is not unlikely to have been one of the spectators.

Captain Eades And His Mermaid

This dried mermaid had been taken to London by an American sea captain named Samuel Barrett Eades. He was by no means a young man and had, for many years, served the well-known Boston commissioning house Perkins & Co. He owned one eight in the merchant vessel *Pickering*, with the London ship owner Stephen Ellery holding the remaining seven eights. While cruising near the East Indies, Captain Eades had the good fortune to save the entire crew from a sinking Dutch man-of-war. Hoping to receive a reward from the government of the Dutch East Indies, the captain set course for Batavia. The negotiations with the Dutch authorities went on for a considerable time, and the adventurous American was soon tempted by novel attractions. Some merchants arranged a private showing of a great curiosity newly caught off Japan: a dried mermaid, skillfully prepared. The Dutchmen said that they had purchased the mermaid from a Japanese fisherman, who had not understood what a great zoological wonder he had caught in his nets. They themselves fully realized its value, and the mermaid was prized at 5,000 Spanish dollars, or 1,200 pounds—a small fortune in these days. Captain Eades was enchanted by this curious specimen, which had no apparent junction between the torso and the fish tail. He did not doubt for a moment that it had once been a Maid of the Ocean. The captain dreamt of buying it for the money he would be given by the resident of Batavia, and then to take it for a triumphant tour to London and New York. Here, he would be cheered by the masses as a great explorer and the discoverer of a novel species of animal, the missing link between the human being and the fishes. To his chagrin, Captain Eades received no reward at all from the Dutch government. Nevertheless, he could not part with the alluring mermaid, and in January 1822 he made the fateful decision to buy it. He obtained the money by selling the ship *Pickering* for 6,000 dollars! The ship owner in London is likely to have received an unpleasant shock when the naive captain sent him a letter telling that he had sold the ship, with cargo and all, without authorization, and that Captain Eades was proceeding toward London as a passenger on board another vessel.

When the ship on which he was traveling temporarily cast anchor in Cape Town, Captain Eades exhibited the mermaid to boost his traveling funds. There was a great rush of interested spectators, many of them full of reverence for this unheard-of *lusus naturae.* A missionary was very much affected by this strange spectacle and called out in ecstasy that he could now bless God and die, for he had beheld a mermaid. Dr. Philip, the representative of the London Missionary Society in Cape Town and a distinguished man of the church, wrote a long letter to the *London Philanthropic Gazette,* which was later reprinted by the *Gentleman's Magazine* and many daily newspapers. It began with the words: "I have to day seen a Mermaid, now exhibiting in this town. I have always treated the existence of this creature as fabulous; but my skepticism is now removed." He then described the mermaid thoroughly. Its head rather resembled that of a baboon, being thinly covered with black hair. The eyes, nose, chin, breasts, fingers, and nails resembled those of a human being. The features were contorted, as with pain or terror. The mermaid was approximately 3 feet long, but Dr. Philip did not doubt that it had shrunk during preservation. He had no doubt that it was a genuine specimen, and his article was an excellent advertisement for Captain Eades, particularly since it was mentioned that the captain was going to London to exhibit the mermaid.

In September 1822, Captain Eades arrived in London. The relentless bureaucracy had an unpleasant surprise in store for him, however, since the mermaid was at once confiscated, as contraband, by some zealous customs officers. It was held for several weeks, while it was investigated whether these creatures were to be subjected to import duty or quarantine. As soon as the captain had managed to liberate his mermaid from the customs officials, he advertised in the London papers for an exhibition theater in which to show it to the public. A certain Mr. Watson, owner of the Turf Coffeehouse in St. James's Street, rented him a large room adjacent to the coffeehouse. They reached an agreement after the captain had agreed to pay a special fee to cover the wear and tear of the carpets. This is likely to have been a clever move by the cunning publican, since the mermaid exhibition was crowded with people already from the start: both the captain and the impre-

A print issued in 1822 to advertise Captain Eades's mermaid. ©The British Museum. Reproduced by kind permission.

sario he had hired were fully employed to guide the visitors and to tell tall tales of her marine ladyship. To advertise the mermaid further, the famous artist George Cruikshank was hired to draw it for an advertisement. His remarkable drawing remains the best illustration of what the specimen really looked like; only a few copies are preserved today, one of them in full color. Captain Eades also put an advertisement in all the major newspapers:

> THE MERMAID!!!—The wonder of the World, the admiration of all ages, the theme of the Philosopher, the Historian, and the Poet. The above surprising natural production may be seen at No. 59, St. James-street, every day, Sundays excepted, from ten in the morning until five in the afternoon. Admittance One Shilling.

The Mermaid in London

Throughout the autumn of 1822, the mermaid was London's greatest scientific sensation: people thronged to see it, and most newspapers had articles about "the remarkable Stuffed Mermaid." It is fortunate that the public were not allowed to touch or handle the mermaid, since it would, in that case, soon have disintegrated. It was jealously guarded by Captain Eades and his assistant and protected by a heavy glass dome, inside which the creature stood erect, supported by its tail. In an article in the *Mirror* newspaper, it was reckoned that every day, three to four hundred people paid their shilling to visit the exhibition. Although the majority of men of science had, by this time, discarded the belief in mermaids as an old wives' tale, the general public was unconvinced, and the belief in the existence of mermaids and other strange sea monsters lived on. In the *Gentleman's Magazine* of October 1822, there was an article by a certain Dr. Rees Price in which Captain Eades's mermaid was accepted as a novel species. Dr. Price had been allowed to take the mermaid out of its glass case and to examine it closely. The head was the size of that of a 2-year-old child, and half of it was covered with black hair resembling that of a human being. The other half had lost its hair, probably through the ravages of time. The jaws were prominent, and the teeth short and sharp. On the abdomen, just below the breasts, were two large fins. Dr. Price was much impressed by the fact that there was no apparent junction between the mermaid's upper body and the scaly fish tail. He concluded that it was now indisputably proved that the old scholars of the sixteenth and seventeenth centuries had been right: mermaids did really exist, and "the introduction of this animal into this country will form an important aera in natural history." Several other naturalists went to see the mermaid, and the majority of them supported Dr. Rees Price. There were several long articles in *Gentleman's Magazine* and other periodicals, comparing it with other specimens, some of them observed several hundred years earlier.

There was certainly no shortage of older observations of mermaids, living or dead. As early as 1403, a living mermaid had been caught off Edam in Holland. It lived on dry land for several years, dressed in woman's apparel and soon learned to spin. Its religious education was not neglected, and it is recorded that although the mermaid did not speak, it always made devout reverences with its tail each time it passed a crucifix. In 1531, another live mermaid was caught in the Baltic and was sent as a present to King Sigismond of Poland, where it lived for 3 days and was seen by all the court. In 1560, some fishing Jesuits at Ceylon caught seven tritons and seven mermaids, some of which were dissected by a doctor who had joined the clerical fishing party. Several early-seventeenth-century mariners, the explorers Henry Hudson and Sir Richard Whitbourne among them, reported mermaid sightings. Another remarkable specimen, the famous mermaid of Amboine, was described in 1718 by the Dutchman Samuel Fallours. It was kept alive in a tub for more than 4 days, uttering cries like those of a mouse. Its portrait was presented to King George III of England, Czar Peter the Great of Russia, and many other dignitaries.

In the seventeenth-century collections of curious medical and zoological case reports, mermaids were often discussed. In his museum, the famous anatomist Thomas Bartholin kept a hand and a tooth of a *Sirena danica*, the Danish mermaid. These specimens were later purchased by the king of Denmark and resided in his museum for more than 150 years, before being sold at public auction in 1826. In 1749, another Danish mermaid was said to have been caught off Nykøping, in Jutland; many naturalists traveled to see it. Carl Linnaeus was very interested in this specimen, and wrote an exulting letter to the Swedish Academy of Sciences, declaring that if this specimen was real, it would signify a major breakthrough in natural history. He himself prepared to go to Denmark to see it, but before he had time to leave Stockholm, he was reached by dispatches that it had been exposed as a fake. It is likely that Linnaeus nevertheless had occasion to see at least one mermaid, since he mentions a specimen of "siren" from Brazil, kept at the museum of Leyden, in the tenth edition of his *Systema Naturae.* A most remarkable mermaid was exhibited at the market of St. Germains, in 1759. It was about 2 feet long, alive and very active, and sported about in a vessel of water. It was fed bread and small fishes. Its features were hideously ugly, the ears huge, and the back part of the body, including the tail, covered with scales. The British Isles also had their fair share of mermaid sightings. In 1737, one had been exhibited at Exeter, and in 1800, two young tritons had been caught near the Isle of Man. There had also been two much-publicized sightings of mermaids off the Scottish coast, in 1809 and 1812.

The Mermaid on Trial

At least one London gentleman was not at all amused by the hullabaloo about the captain and his mermaid: the ship owner Stephen Ellery, whose ship Captain Eades had sold, without authorization, to buy the mermaid. Mr. Ellery visited the exhibition to reproach the swindling captain and demand his money back, but Samuel Barrett Eades was unwilling to reimburse him, or even to give him a fair share of the considerable profits from the exhibition. He threatened that if Mr. Ellery tried to take him to court, he would leave the country, taking his mermaid with him. Mr. Ellery was a cunning businessman, however, and he thought of a clever way to prevent Eades from moving abroad. In the early 1800s, it was of frequent occurrence that adventurers abducted wealthy heiresses, whom they had previously seduced, to marry them without the consent of their parents. To stop these immoral practices, the parents could appeal to the lord chancellor's court, since the lord chancellor had the authority to make an eloped young lady his ward (a ward in Chancery); she was then not allowed to marry without his permission. On November 20, 1822, Mr. Ellery appeared before the court of Chancery to restrain Captain Eades from moving or selling the mermaid. His solicitor Mr. Hart delivered a long speech about the deceitful captain and the circumstances of his purchase of the mermaid. It is recorded that Lord Eldon, the lord chancellor, listened to his harangues with some mirth. The main point was whether the object—"be it man, woman, or mermaid"—had been purchased from the produce of the ship *Pickering*. Mr. Hart swore that this was the case, and read an affidavit that Mr. Ellery was the ship's main owner. The lord chancellor then decided that Captain Eades was to be restrained from removing or disposing of the mermaid without the lord chancellor's permission.

It bears repeating that Captain Eades himself seems to have remained, throughout all these strange proceedings, entirely convinced that his mermaid was a genuine specimen. As soon as he had come to London, he had arranged that the doyen of London anatomists, Sir Everard Home, was to inspect it and make an affidavit of its origins as a Maid of the Ocean. They agreed that Home would have the sole rights of describing the mermaid in some scholarly journal; in exchange, he had to promise not to write to the newspapers or gossip about his findings to other interested parties. As we know, the mermaid was seized as contraband, and it took several days for Eades to get it back from the customs officials. Sir Everard was busy at the time, and he ordered his assistant William Clift, the Conservator of the Hunterian Museum, to inspect it instead. On September 21, Clift went to the East India Baggage warehouse, where the mermaid was still held by customs officials. Captain Eades waited for him there and reverently took the mer-

Subjoined is a sketch of the front and back view of the object, which give a general Idea of its form and proportions; and though drawn from recollection (immediately after having examined it) has a greater resemblance than any of the figures which have been published; none of which, it is evident have been drawn from the figure itself.

Nov.r 29. 1822.

W. C.

William Clift's description and drawing of Captain Eades's mermaid, from his manuscript account, kept in the archives of the Royal College of Surgeons (Cabinet II.7). Reproduced with kind permission of the President and Council, Royal College of Surgeons of England.

maid out of its tin case and unwrapped a silken mattress that had protected it from injury. William Clift, who was an expert anatomist and practical zoologist, then examined the specimen minutely. It was soon apparent to him that it was a palpable imposition, and he could also tell how it had been manufactured. The mermaid was exactly 2 feet 10 inches long, of a mummy-like brown color. The cranium and torso belonged to a female orangutan of full growth, but the jaws and teeth were those of a large baboon. The scalp was thinly covered with the orangutan's own hair, but the skin of the face had been artificially joined to the skin of the head across the eyes and upper part of the nose; the nose and ears had been constructed from folds of this skin. The eyes were artificial. The bones of each lower arm had been sawed through, beneath the skin, to bring them closer to the proportions of human extremities; although this had been skillfully done, Clift could feel the sharp edges of the sawed bone. The nails had been made of horn or quill. The breasts had some slight stuffing, and below them a deep fold of skin hid the junction with the fish tail. From the number of fins and their position, Clift suspected that the fish part of the mermaid was the entire body of a large specimen of a fish belonging to the salmon genus, separated from the head immediately behind the gills. William Clift grudgingly had to acknowledge the skillful way in which the fish's skin had been preserved as high as possible toward the head and then pressed down upon the spinous processes of the orangutan's dorsal vertebrae, to give the impression that the spinal column of the ape continued farther down the fish tail. There was no apparent junction between the two parts of the specimen, but Clift could feel that a hoop of some firm substance had been used to distend the body of the fish from the pectoral to the anal fins; below this hoop, the fish tail had shrunk considerably, leaving a distinct edge all around.

William Clift wrote a detailed account of his findings and reported them to Sir Everard Home, who probably saw the specimen himself at some later date. It is likely that Captain Eades received an unpleasant shock when the two anatomists told him, in so many words, that his mermaid was nothing but a fake; he must have blessed his precaution in swearing them to silence, which enabled him to keep exhibiting it. He later, as we know, consulted several other, less skilled naturalists, who accepted the mermaid as a genuine specimen. This apparently did much to restore the captain's faith in his mermaid. In November 1822, at the height of the mermaid hysteria in London, he was seized with a fateful attack of hubris. He was unwise enough to advertise, both in the newspapers and on the signposts, that Sir Everard Home had declared the mermaid to be genuine. Home was enraged by this low trick. He considered his previous agreement with Eades to be broken and ordered William Clift to write to the newspapers to denounce the mermaid and its unscrupulous owner. Clift's long and witty article was published in *the Morning Herald* and was later covered in several other newspapers: it heralded the beginning of the end of Captain Eades's high hopes for the mermaid.

The Mermaid Revealed

In the December issue of the *Gentleman's Magazine,* a publication hitherto favorably disposed to the mermaid, a certain Mr. J. Murray, F. L. S., delivered another broadside against this unfortunate creature and its owner. Although Mr. Murray was not surprised that the ignorant London populace had thronged to see the mermaid, he was astonished that many men of observation had made out affidavits of its genuineness. In particular, Mr. Murray sneered at Dr. Philip and Dr. Rees Price for not being able to see through such a palpable imposition. For any person that kept abreast of the scientific development, it was impossible to imagine a creature being half fish and half mammal. Dr. Morrison, a much-traveled friend of Mr. Murray, supplied some lurid anecdotes about the habits of the yellow tribes who, it was alleged, were keen on fabricating faked mermaids to deceive unsuspecting Europeans. Sometimes, young children were kidnapped for the purpose of turning them into the upper part of a mermaid. According to the macabre Dr. Morrison, this outrage occurred frequently, and the children were, like the orangutans, cruelly tortured to make their features hideous and distorted: "so much for the *Empire* of the *Celestials*, and the character of that mild and harmless *Religion*, which some persons, even in *Christendom*, affect to advocate." Finally, Mr. Murray suggested that, since the damsel was now a "Ward in Chancery," the Lord Chancellor should order a London anatomist of repute to dissect the specimen, to find out "whether this 'Mermaid' is, what it purports to be, a Maid of the Ocean."

In spite of these setbacks, Captain Eades and his friends continued to exhibit the mermaid, but it had now been in the metropolis for more than 4 months, and the fickle public was tempted by novel amusements. The scholarly debate in the *Gentleman's Magazine*, in which the supporters of the mermaid were suffering ignominious defeat, was covered in most daily newspapers. As the Londoners were reading the sneering remarks about the ignorant and stupid populace being imposed on by the mermaid and its owner, they naturally became increasingly unwilling to visit the exhibition. In December 1822, a longer exhibition pamphlet, entitled *A Historical Memoir on the Existence of the Marine Animals, Called Syrens, or Mermaids*, appeared. It had probably been commissioned by Captain Eades. It is evident that it was written by a competent naturalist, since the pamphlet contains a scholarly review of mermaid observations from the fifteenth century onwards. An extract was later printed in the *Gentleman's Magazine* as an addendum to an article by one of the mermaid's last defenders. The actual specimen is but briefly described, and the writer wisely leaves it to the reader to determine its genuineness. This pamphlet did not have the desired effect, however, and the mermaid exhibition had definitely lost the public's favor. In late December, Captain Eades had to declare that although the mermaid was now a Ward in Chancery, it was leaving the country in a few weeks; the members of the nobility and gentry still wanting to see it thus had to hurry up. To celebrate the mermaid's unique status as a Ward in Chancery, and to mock the Lord Chancellor, George Cruikshank made a drawing entitled "A Mermaid in Chancery Holding in Tail," depicting Lord Eldon holding his ward in a tender embrace. It may well be that this drawing had also been commissioned by Captain Eades. Through a spy, the Lord Chancellor found out about their plans, however, and the print made from this drawing was sup-

George Cruikshank's drawing intended for the caricature print *A Mermaid in Chancery Holding in Tail*. Reproduced with permission of the trustees of the Victoria & Albert Museum, V & A Picture Library.

pressed. Not a single copy exists today, but Cruikshank's original drawing is still kept at the Victoria & Albert Museum.

In spite of Captain Eades's advertising talents, public interest in mermaids continued to be low, and on January 9, 1823, the Turf Coffeehouse exhibition was unceremoniously closed down. For several months subsequently, the mermaid had to suffer the ignominy of being exhibited with another artist, and even being underbilled to this more respected performer on posters and advertisements: it was Toby, the Learned Pig, who stole the show from her marine ladyship, both at Bartholomew Fair and at the Horse Fair. According to a newspaper account, the manager of the show was once visited by two muscular thugs who demanded entry without being able, or willing, to pay. After they had been thrown out by the bouncer, they swore to return the next day, to beat up the showman, cut the mermaid in two, and barbecue the learned pig for their dinner. The exhibitor lost his nerve faced with this dire threat to his business: he rushed to the Marlborough Street Court, with the mermaid under his arm and Toby on a leash. There, he begged the judge that a constable should be put on guard outside the showroom, to protect Toby and the mermaid. This incident was widely publicized in the papers, but there were no further hostilities at the showroom. In fact, the exhibitor of the mermaid gained from this incident, since many people frequented his rundown establishment, hoping to witness fights and murder.

During the autumn of 1823, the mermaid toured the provinces, but with less media attention than during its heyday; it had lost its novelty, and newspapers all around Britain had published articles claiming that it was a fake. The only newsworthy incident on this tour happened in County Cork, where the mermaid was exhibited at the museum of a certain Mr. Lefevre; it was recorded under the title "A Novel Case of Abduction." While Mr. Lefevre was lecturing, a thief suddenly seized the mermaid and rushed downstairs. He was fortunately arrested while running out from the museum, and "the *fair* prize restored to her home. It is doubted whether the delinquent may not be brought before a very high Tribunal for this offence, as the *Maid* is stated to be a Ward of Chancery."

The mermaid's tour of the provinces continued well into 1824, and it is likely that her marine ladyship was back at Bartholomew Fair in 1825. Here, she was exhibited—without her porcine companion—by a clever and enterprising showman. He put up a large sign with the portrait of a ravishingly beautiful mermaid girl outside, which helped business considerably, although it is surprising that no enraged Londoner used his cane to knock the head off the shriveled, hideous specimen grinning at their astonished faces from within its glass dome. From 1825 to 1842, the mermaid's whereabouts are unknown. It may be that it was back in London in the summer of 1836, to be exhibited together with a learned pig in a twopenny show, but this remains unproven, since no drawing of the mermaid was published at the time. William Clift claimed to have seen the mermaid in a French country museum in 1838, and this may well be true, since he was a trained observer with a retentive memory. Another mermaid, which was said to have been imported from England, was present at Mr. Dorfeuille's Cincinnati Western Museum in the 1830s. There were several faked mermaids about during this time, however, and, in the absence of an illustration, it is difficult to keep them apart.

P. T. Barnum and the Feejee Mermaid

After nearly 20 years in obscurity, the mermaid once more shot to fame in 1842, on the other side of the Atlantic. Mr. Moses Kimball, the proprietor of the Boston Museum, traveled to New York to meet his friend, the showman Phineas Taylor Barnum, who was well known for his cleverness and talent for advertising. Moses Kimball suggested that they cooperate to earn big money on a great curiosity he had recently bought from a Boston seaman, a marvelously well preserved dried mermaid. The seaman had told him that the mermaid was once the property of his deceased father, who had been a sea captain. The captain had bought the mermaid in the Far East, by selling his ship for 6,000 dollars and misappropriating the money. After exhibiting the mermaid for several years, Captain Eades (for it must have been he) was

Phineas Taylor Barnum, at the time he acquired the Feejee Mermaid.

taken to court by the shipowner and sentenced to pay back the money he had embezzled. The destitute captain returned to his old job and had to serve the shipowner for many years, probably never entirely repaying his enormous debt. Like a bizarre Flying Dutchman, he sailed the high seas for nearly 20 years; although he did not carry the mermaid tied around his neck, he probably joined Coleridge's Ancient Mariner in cursing his hard fate and the creature that had caused his ruin.

After the old captain's death, in the early 1840s, the only thing bequeathed to his son was the mermaid. Mr. Eades Jr. exhibited it in his house, before his amazed friends. Moses Kimball heard of the mermaid's existence and made young Eades a modest offer for it; the sailor did not hesitate to turn his strange family heirloom into cash. Although P. T. Barnum was also quite impressed with the mermaid's apparent authenticity, he did not take Kimball's word for granted. The cautious showman took the mermaid with him to an American naturalist, to consult him about its genuineness. The naturalist at once pronounced it to be a manufactured specimen. When Barnum asked him why, he replied "Because I do not believe in mermaids." P. T. Barnum had been long enough in his particular line of business, however, to know that naturalists were no more than ordinary men, with their petty feuds and disputes; if one of them decided not to believe in mermaids, others would take the opposite stance. He himself believed in mermaids, or at least their ability to take in his more gullible countrymen, whom he knew to have an eager appetite for curiosities. Some days later, he made a deal with Moses Kimball, leasing the mermaid for 12 dollars and 50 cents a week and agreeing to hire it a manager, whose salary was to be paid by Barnum.

P. T. Barnum was, at this time, 32 years old and at the beginning of his long and illustrious career as America's self-confessed "Prince of Humbugs." Already in 1835, he had become notorious by exhibiting the aged Negress Joice Heth, alleged to be 161 years old. During the shows, she smoked her pipe, sang ancient hymns, and told the audience about the habits of young George Washington, whom she had claimed to have served as a wet nurse. Barnum had made a deal with an eminent New York surgeon, Dr. David L. Rogers, that he be allowed to dissect this supercentenarian when she died. When poor Joice expired in early 1836, the doctor was dismayed to see that it was only too apparent that she could be at most 80 years old! He withdrew from the proceedings in a huff, but Barnum and his friend Levi Lyman had the last laugh: they started a rumor that Joice Heth was still alive, and that the body was that of another woman. This version was widely believed by the journalists.

Throughout the 1830s, Barnum continued as an itinerant showman, before buying Scudder's Museum in 1841. In his autobiography, Barnum remarked that the atmosphere of this New York establishment was about as dead and crumbling as the skeletons and stuffed animals exhibited therein. Barnum spent hundreds of dollars on advertising his museum, but he needed a great attraction to publicize it in a big way.

As soon as the mermaid had come into his hands, Barnum's fertile mind assessed the situation, and he thought of a masterly plan for cashing in on his new acquisition. After meticulous preparation, a series of letters was sent to various newspapers, telling that the famous British naturalist Dr. Griffin, of the (nonexistent) London Lyceum of Natural History, was coming to the United States. The doctor brought with him a great curiosity—a mermaid caught off the Feejee Islands—and it was hoped that he could be prevailed on to show it to the public. Some days later, Barnum's trusted assistant, the clever lawyer Levi Lyman, who had been hired as the mermaid's manager, registered at a Philadelphia hotel as Dr. Griffin, of Pernambuco and London. When learning who he was, the landlord and a few of his friends, some of them newspaper men, were reluctantly given a brief audience with the Feejee Mermaid, as the specimen was now called.

The day after, when Dr. Griffin left for New York, the Philadelphia papers were full of the mermaid story. When Dr. Griffin registered at the Pacific Hotel in New York, reporters were waiting for him. Again, with seeming reluctance, the doctor let the panting newshounds view his mermaid. Shortly after, P. T. Barnum himself contacted the editors of three major New York papers, telling each of them that he had, in vain, tried to persuade Dr. Griffin to let him exhibit the Feejee Mermaid. The doctor had once more been reluctant to exhibit it before the American public, and even Barnum had to admit defeat. Magnanimously, he offered each editor the sole rights to a drawing of the mermaid, which he had ordered at great cost. Each editor accepted this offer with alacrity, but when the articles were printed on the following Sunday, the newspaper men realized that they had been taken for fools: Barnum had released different mermaid prints to each newspaper. The same day, Barnum had ten thousand leaflets distributed, telling that the stubborn Dr. Griffin finally had given in to popular opinion: the mermaid would be exhibited, but only for 1 week. Barnum rented the grandiose Concert Hall at Broadway, which was thronged by curious Americans. The fraudulent Dr. Griffin held well-attended lectures about his exciting life as an explorer, his travels to the Far East, and his singular zoological theories. He was a pre-Darwinist, and believed in the Great Chain of Being. Just as

One of Barnum's drawings of the Feejee Mermaid.

the mermaid was the missing link between humans and fishes, the flying fish connected the birds and the fishes, the South American mud iguana the reptiles and the fishes, and the Australian duck-billed platypus the seal and the duck. Another naive argument was that since there were sea horses, sea lions, sea elephants, and sea dogs, there was no reason to suppose that the human being should have no subaqueous counterpart! This reasoning was echoed in a pamphlet entitled *A Short History of Mermaids*, which had been compiled by Barnum himself, according to his autobiography. Large parts of this booklet had been reprinted from the 1822 pamphlet commissioned by Captain Eades; his son might have delivered a copy to Moses Kimball along with the specimen.

At least at the beginning, Levi Lyman must have had an amusing time as the Feejee Mermaid's manager. He pulled people's legs by the thousand, and had delivered his harangues about mermaids, sea dogs, and the Great Chain of Being with a perfectly straight face. He himself was once the subject of ridicule, however. Some waggish medical students came to see the Feejee Mermaid, and when Lyman once turned his back for a moment, one of them forced open the mermaid's glass case and thrust a long, lit cigar between the figure's jaws! Levi Lyman did not notice this outrage: he started his customary lecture, recounting old legends of mermaids before describing, in some detail, how his own specimen had been caught off the Feejee Islands. One of the students then asked whether the mermaid had, at the time, been smoking the same cigar she was now puffing on inside her cabinet! Barnum later remarked that this was the only time he ever saw the ready-witted Lyman completely at a loss for words.

After the triumphant week at the Concert Hall, Barnum removed the mermaid to his own American Museum. The takings of this establishment tripled as soon as the mermaid had been taken there, and Bennett's Museum, a rival cabinet of curiosities, saw no other course of action but to order its own mermaid from one of New York's most skilled taxidermists. The main part of the profits from the Feejee Mermaid's exhibition went into Barnum's own pockets, which seems to have annoyed Levi Lyman. He was the one who had to answer to the many indignant and foul-mouthed Yankees who were complaining that the shriveled, hideous specimen did not look much like their idea of a mermaid. From Barnum's advertising, they had expected a live mermaid, or at least something with a nude female torso, and it was hard work for Levi Lyman to appease them. When Barnum planned to hang up an enormous flag, depicting an 18-foot, alluring mermaid, Lyman objected strongly. Barnum had to give in, and the flag was discarded. Nevertheless, Levi Lyman had had enough of the mermaid, and he left the American Museum within a couple of months; he later went to Illinois, where he became one of America's most influential Mormons.

In the summer of 1843, when the New Yorkers had finally begun to tire of the Feejee Mermaid, P. T. Barnum sent it for a tour of the southern states, under the management of his uncle, Alanson Taylor. The tour initially went smoothly, but Taylor, a former schoolmaster, wholly lacked Levi Lyman's talent for imposition and later got himself and the Feejee Mermaid deeply into trouble. In Charleston, he became involved in a long-standing feud between two local newspapers. Whereas the editor of the *Charleston Courier* was a supporter of the mermaid's genuineness, the journalists of the *Charleston Mercury* attacked the Feejee Mermaid and its exhibitor with great vehemence. A clergyman and amateur naturalist, the Reverend John Bachman, consulted a committee of scientists from the University of Charleston, led by the professor of mathematics, Lewis R. Gibbes, and the professor of anatomy, J. Edwards Holbrook. They unanimously declared that the Feejee Mermaid was nothing but a fake. The credulity of the Charleston public was mocked by the newspaper writers, in deeply insulting terms. This treatment angered many previous visitors to the exhibition, and poor Taylor was harassed and manhandled by the mob. P. T. Barnum suggested that Taylor and Yeardon should sue their opponents to gain further publicity; the Feejee Mermaid's tour would then continue abroad. Barnum knew a popular Universalist minister, the Reverend Edward Chapin, who agreed to make a sermon about the Feejee Mermaid, but his harangues about the failure of smaller souls to fathom God's limitless creativity and the myriad links of beings did nothing to appease the enraged Charlestonians. Although Barnum tried to spur him on, the wretched Alanson Taylor had had enough of the showman's life for good; according to Barnum, "poor Taylor has gone through anything a mortal could stand." The Feejee Mermaid was hurriedly sent back to New York to prevent its destruction by the individuals who were threatening to attack the exhibition.

Where is the Feejee Mermaid Today?

After spending a couple of years at Barnum's museum, the Feejee Mermaid was returned to Moses Kimball's Boston Museum. When Barnum went on a tour to London in 1859, he took the mermaid with him: it was shown during his lectures and attracted much interest, just as it had 37 years earlier. An elderly visitor gave Barnum an extract from the *Mirror*, of November 9, 1822, containing a sketch of the specimen; this interested the showman greatly, since he had previously suspected that Kimball had exaggerated the details of the Feejee Mermaid's London career. In June 1859, Barnum returned the mermaid to Kimball, and it probably remained at his museum for many years. P. T. Barnum again exhibited a dried mermaid at his circus during the early 1880s. It may have been the Feejee Mermaid, but it should be noted that, according to his account books, Barnum ordered a newly made mermaid from

well, since the mermaids were again sold at auction 2 months later. According to Mr. A. H. Saxon, the biographer of P. T. Barnum, none of these three specimens resembles the real Feejee Mermaid, nor does any other American mermaid he has seen. I have inspected many European specimens with the same disappointing result. It is thus unlikely that the Feejee Mermaid still exists today; it may be that it was burned in one of the fires mentioned previously, or that it was destroyed by some individual who did not realize its value to posterity.

Other Faked Mermaids

The manufacture of faked mermaids can be traced far back in time. According to Purchas his *Pilgrimage*, a "Mermaid skinne" was on show at Thora, a busy port at the Red Sea, in 1565. It is also recorded that in 1660 a dried mermaid was hanging in the church of Swartvale, Holland. The first mermaid to tour Britain appeared in 1737: it differed from the majority of its successors by having wings, webbed feet, and a blow hole. This strange amphibian mermaid was shown in London and Exeter. When Mr. Murray attacked Captain Eades's mermaid in the *Gentleman's Magazine*, one of his arguments was that he had seen similar specimens exhibited in 1775 and 1812. The one touring Britain in 1774 and 1775 was, according to several witnesses, particularly well constructed. It was claimed in the newspapers that only one superior specimen existed—one that was 226 years old, belonging to the collection of an archduke of Austria. During its tour of Britain from 1774 to 1775, the mermaid was seen and admired by the politician and philosopher Edmund Burke, who did not doubt its authenticity. A mermaid appearing in London in 1809 was less favorably received. A naturalist declared it to have been manufactured from the skin of a shark, and its owner, who had previously exhibited it at fairs and markets with considerable success, was arrested, charged with fraud, and fined a large sum of money.

The exhibition of faked mermaids in London continued long after the Feejee Mermaid had been taken to the United States. Frank Buckland had occasion to examine not less than three alleged dried mermaids during the 1850s: one of these was advertised for sale at Messrs. Farmer & Rogers Oriental Warehouse, while another—a particularly hideous merman—was exhibited in the back parlor of the White Hart, Vine Court, Spitalfields. Frank Buckland particularly liked a small mermaid specimen for sale in the old curiosity shop at Hungerford Market. All these were manufactured in the same way as the Feejee Mermaid: the upper body of a monkey had been skillfully joined to the hind part of a large fish, usually a hake, salmon, or carp. The junction was hidden, and the fur and the scales seemed to join in a natural way. It may well be that Frank Buckland purchased one of these mermaids, since in his will he left a mermaid specimen to the Museum of the Royal College of Surgeons in London, which already contained another one; both of these were casualties of a German bombing raid in 1941.

In 1878, a dried mermaid was put on show in Ostend, and as late as 1881, another specimen was exhibited in New Orleans. In certain Mediterranean harbors, the practice of keeping these creatures at hotels and bars continued even longer. In the 1940s, the American writer Peter Lum saw a sign stating "Mermaid on View" at Aden. He unfortunately does not give any details of the specimen, nor does the famous author Evelyn Waugh, who mentions that two Aden hotels boasted mermaids at about this time. When, in 1959, Waugh again visited one of these hotels in Aden, he requested to see the mermaid, but no one seemed to know anything about it. One hotel servant looked blank and shrugged, "supposing I was demanding some exotic drink. But a much older man came forward. 'Mermaid finish', he said." Waugh deplored, in his book *A Tourist in Africa*, that this elderly servant did not know enough English enough to explain exactly what had happened to it except that "One man come finish mermaid"; was it stolen, thrown out as an imposition, or destroyed by a drunk?

Even in the enlightened 1990s, faked mermaids are by no means in danger of becoming an extinct species. In 1961, a mermaid belonging to the zoologist Sir Alister Hardy, F.R.S., was exhibited at the British Museum in London. The mermaid had been presented to him some years earlier by a complete stranger, who evidently believed it to be genuine; it had been in his family for generations, and was believed to be 300 years old. It was a comparatively attractive specimen, with an elegantly curved tail and large glass eyes. When Sir Alister Hardy had it radiographed, the mermaid was found to contain an elaborate system of wires. In August 1990, I myself saw two specimens at an exhibition of faked objets d'art and natural history specimens at the British Museum. One of these is the property of the Horniman Museum, which is a part of the Wellcome Collection. In 1993, I had occasion to reexamine the other specimen, at the British Museum stores in North London. In the late nineteenth century, this mermaid had been given to Prince Arthur of Connaught by a Japanese nobleman; at least the Japanese is stated to have accepted it as genuine. This specimen is likely to be one of the most skillfully constructed mermaids extant today, although it cannot be

The faked mermaid from The Horniman Museum London, Wellcome Collection.

compared with the Feejee Mermaid. It rather resembles this famous specimen in its build, although it is considerably smaller; even the elegantly curved tail is the same. Like most other faked mermaids, it is built from an ape and a fish. A radiograph of the mermaid shows that the fish's backbone was cut off in the area of the figure's navel; patches of fish skin can be found on both forearms, and the jaw is also likely to be that of a fish. The tail is stabilized by two metal rods, and the entire figure is filled with clay or plaster. The head does not contain a bony skull and is probably not solid. The skin has a dried-up, polished appearance, probably from the many hands touching it during its at least 100 years of existence: the British Museum mermaid is probably unique within its genus in having both mustaches and a beard!

Another well-constructed mermaid is in the Royal Pavilion Museum of Brighton: it was donated to this museum by Mr. W. H. Willett, Esq., a well-known collector. I saw another specimen exhibited in Croydon, in 1995, at the Monsters and Miracles exhibition arranged by the *Fortean Times;* it was on loan from the Royal Museums of Scotland, and stated to be of Japanese origin. In addition to the two mermaids in the Peabody Museum at Harvard University, there are several others specimens in various American muse-

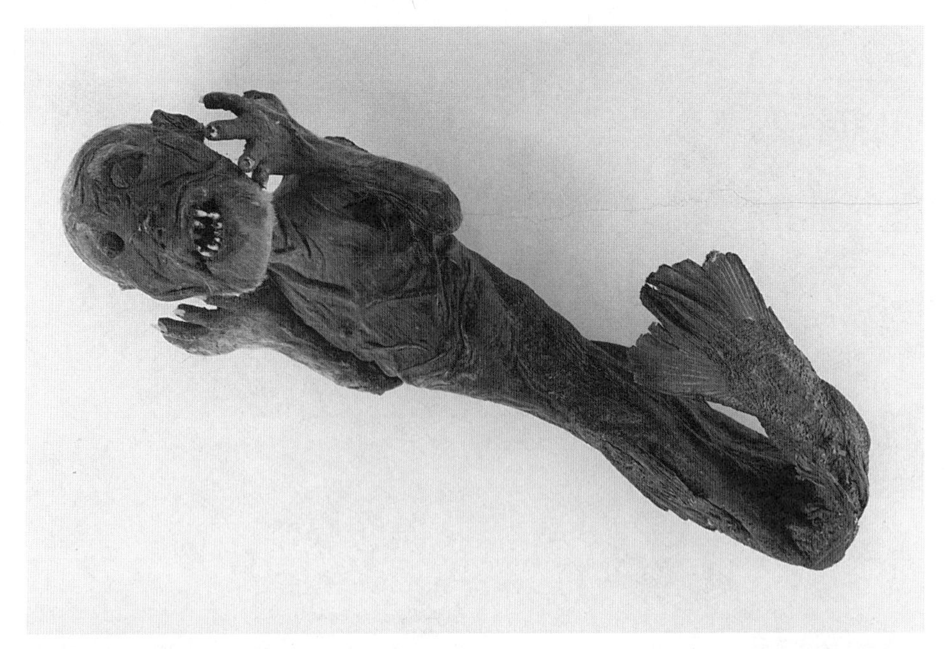

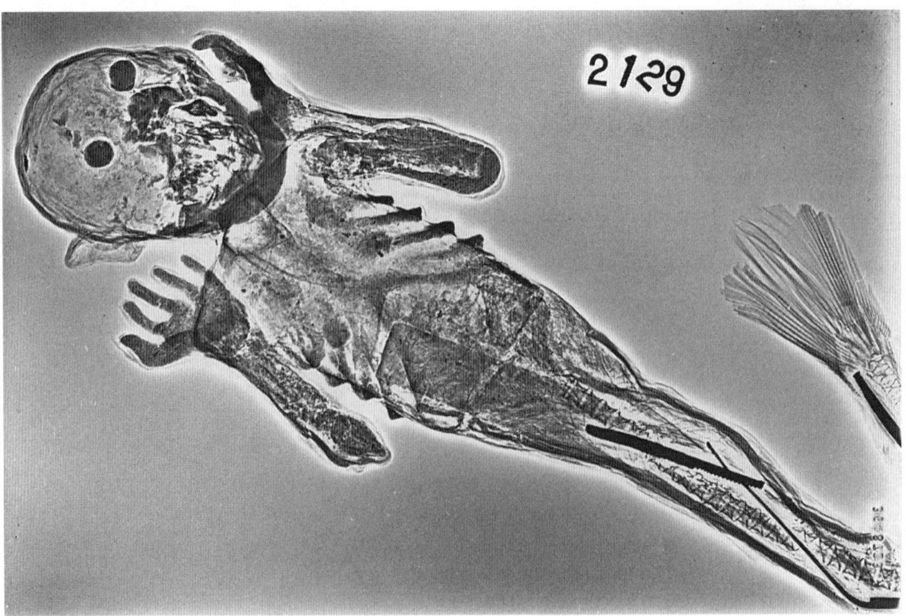

The British Museum mermaid (*A*), with a radiograph showing its construction (*B*). Reproduced with permission of the Museum of Mankind.

ums: a particularly hideous specimen is at "Ye Olde Curiosity Shop" in Alaskan Way, Seattle. "Jake the Alligatorman," who is unique in having the rear part of an alligator instead of a fish, was exhibited at sideshows for many years before coming to rest at Marsh's Free Museum in Long Beach. Holland is another haven for faked mermaids, and quite a few specimens are still kept at various museums. Photographs of several French faked mermaids were published in an article by the teratologist Jean Boullet. An interesting German specimen was donated by Dr. Eugen Holländer to the Kaiserin-Friedrich-Haus: it may well be unique in that its upper part is that of a dried human fetus.

What, then, was the origin of all these faked mermaid figures? Although large parts of the stories told about them by their owners were fabricated, one theme recurs: the origin of the mermaid specimens was Japan or the East Indies. They were constructed by skilled craftsmen or fishermen; often, the ape and fish parts were joined on top of a carefully hidden wooden frame. According to Andrew Steinmetz's *Japan and Her People*, published in 1859, it

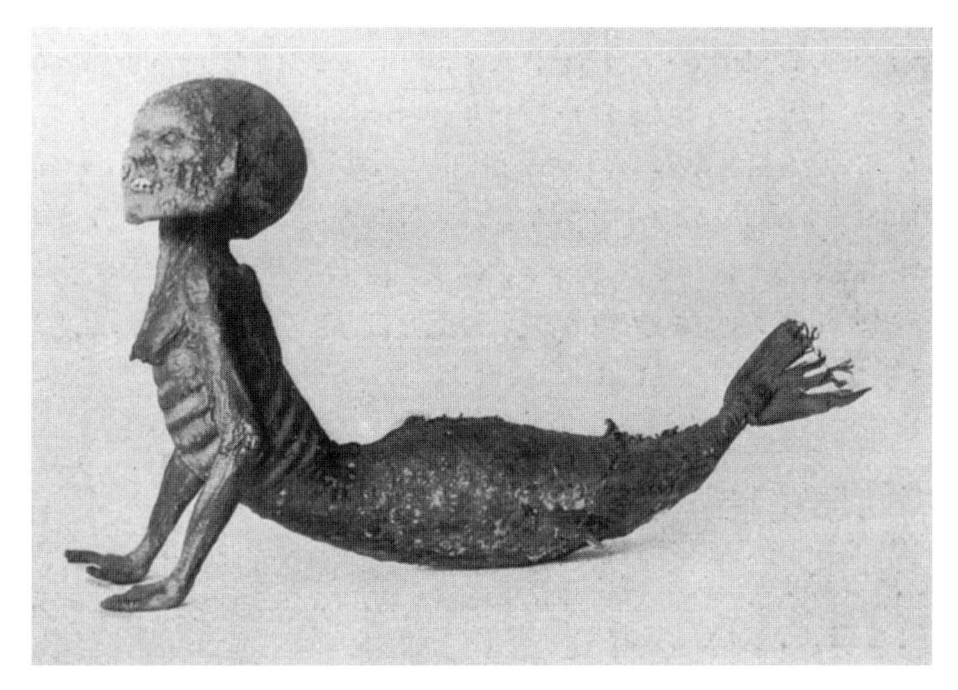

A German mermaid with the front part taken from a human fetus, from the book *Wunder, Wundergeburt und Wundergestalt*, by Eugen Holländer.

was a tradition in these parts to construct faked mermaid figures, dragons, and monstrosities. The specimens were exhibited for money or used in religious ceremonies, and amulets with their likenesses imprinted were sold. When the westerners came to Japan, the trade in faked mermaids was greatly stimulated. Although the bizarre, sphinx-like specimens were quite unlike the usual image of a mermaid, they were often extremely skillfully made, and many travelers were taken in by them. The German ethnologist Ph. F. von Siebold even claimed to have traced the ultimate origin of the Feejee Mermaid, in his *Manners and Customs of the Japanese in the Nineteenth Century.* Some time in the 1810s, a Japanese fisherman had manufactured a particularly detailed mermaid specimen, which he said he had caught alive. Before it had expired, the half-human fish had predicted a certain number of years of wonderful fertility, but then a fatal epidemic would ravage the land. The only remedy for this plague was the possession of the likeness of the marine prophetess. These amulets, which had been made beforehand by the cunning fisherman, had an immense sale. Later, this mermaid was sold to the Dutch factory and transmitted to Batavia. There was a legend in these parts that it had been bought by a speculating American, who had brought it to Europe, exhibiting it in the years 1822 to 1823 "to the admiration of the ignorant, the perplexity of the learned, and the filling of his own purse."

Meddling with Sirens

Some modern ethnologists and historians of science have marveled at the prevalence and longevity of the popular belief in mermaids. Well into the nineteenth century, the myth of water-living, humanoid mermaids remained widespread throughout Europe. It has been presumed that people were influenced by the romantic old legends about mermaids, or by mistaken mermaid sightings by unreliable mariners, but this is unlikely to be the full explanation. Modern zoologists believe that many of the seventeenth- and eighteenth-century traditional mermaid observations represent early sightings of dugongs or manatees. Much evidence supports this notion. It is likely, for example, that the mermaid of the Isle of Amboine was a young dugong. The only objection to this theory is that it would take a half-delirious sailor to mistake the features, or rather lack of features, of the dugong with the face of a beautiful mermaid, or to hear the siren's song in the snuffling sounds made by these ungainly beasts.

Throughout the eighteenth and nineteenth centuries, there was a lively trade in faked mermaids, which were exhibited widely in Europe and the United States. The Feejee Mermaid is likely to have been the most skillfully constructed of them all. Considering the extent of knowledge in natural history in the early 1800s, it is no surprise that a mermaid like this one could be accepted by medical men and zoologists. If scholars like Dr. Philip and Dr. Rees Price were unable to see through the imposition, it is easy to imagine the mermaid's effect on simple, uneducated individuals: thousands of people must have left the exhibition convinced that it was real, and that mermaids really did exist. It is indeed not unlikely that the widespread exhibition of faked specimens contributed to the durability of the mermaid legend.

The exhibition of faked mermaids was a lucrative, if sometimes hazardous, business. Not infrequently, the specimens were bought or sold for considerable sums. The mermaids touring Britain in 1784, 1796, and 1812 were probably the same specimen, since both the illustrations and the advertising texts agree suspiciously well. It is recorded that the mermaid was a reliable "cash cow" during all three tours. P. T. Barnum and Moses Kimball also gained much money and publicity from their involvement with the Feejee Mermaid. But, as in the old legends, there were also men who were ruined by the false, unreliable sirens. In the 1830s, a faked mermaid was shown at the Egyptian Hall in London for several years, before it was sold to two wealthy Italians brothers for 40,000 dollars. These gentlemen later regretted their purchase, which led to a bitter legal feud, in the course of which the mermaid was demonstrated to be a fake. The outcome was that the Italians were completely ruined by their ill-judged investment.

The individual to lose most from his transactions with the false sirens, however, was the wretched Captain Samuel Barrett Eades. Like the seamen in Lord Tennyson's *The Sea Fairies*, the Boston mariner had become entangled by the mermaid's magic; like Sir Guyon in Spenser's *Fairie Queen*, he was led astray onto paths "where many mermaids haunt, making false melodies." In spite of his many setbacks, it is said that Eades's belief in the mermaid's genuineness remained unchanged. During the 20 years he had to sail the high seas without pay, to repay the debt caused by his purchase of the treacherous mermaid, he would have had ample time to consider the following advice from Samuel Butler's *Hudibras:*

Alas! what perils do environ
That man who meddles with a siren!

Obituary for an Elephant

Natures great master-piece, an Elephant,
The onely harmelesse great thing; the giant
Of beasts; who thought none had, to make him
wise,
But to be just, and thankful, loth t'offend
(Yet nature hath given him no knees to bend)
Himself he up-props, on himself relies,
And foe to none; suspects no enemies,
Still sleeping stood; vex't not his fantasie
Black dreams, like an unbent bow, carelesly
His sinewy Proboscis did remisly lie.

In which as in a gallery this mouse
Walk'd and survey'd the rooms of this vast house,
And to the brain, the soul's bed chamber, went,
And gnaw'd the life cords there; like a whole town
Clean undermin'd the slain beast tumbled down;
With him the murth'rer dies, whom envy sent
To kill, not scape; for onely he that meant
To die, did ever kill a man of better roome;
And thus he made his foe, his prey and tombe:
Who cares not to turn back, may any whither
come.

John Donne, *The Progresse of the Soule*

ELEPHANTS HAVE, SINCE ANCIENT TIMES, BEEN used in the service of humankind. They have not only been employed in war and for heavy labor but also used for show and entertainment. King Assurnasirpal II of Assyria, living in the ninth century B.C., kept a zoological garden that contained, among many other animals, several elephants caught in Syria. The elephants entered the Western world with Alexander the Great, who brought several captured war elephants back with him as spoils of war; they became the pride of the Macedonian court. According to Pliny, the Romans saw elephants for the first time in the war against King Pyrrhus, calling them Lucan oxes because they were first seen in the province of Lucania. The noble Romans often used their elephants in ceremonies and victory parades. From the time of the Augustus onward, it was customary for the emperor to travel in a chariot pulled by four elephants during processions and triumphal festivities. In the gory animal baitings at the amphitheater, elephants fought bulls or each other. When, during the second consulate of Pompey, a great entertainment was enacted, not fewer than twenty elephants fought a company of gladiators. The spectators considered it "most amusing" to see the great beasts fling the shattered gladiators up into the air with their powerful trunks. In the times of Claudius and Nero, the final battle in the arena was always between an elephant and a well-armed gladiator.

The Roman war elephants were trained to tread the enemy soldiers underfoot, or to throw heavy arrows at them with the trunks. In contrast, the ceremonial elephants were particularly docile. These beasts were trained to endure clamorous and noisy crowds without bolting, an ability that was tested by keepers, who sneaked up near the elephants during the solemn processions to blow bugles or strike cymbals without warning. The rhetorician Ælian described how, on one occasion, twelve of the most accomplished elephants were marching in circle in a theater, moving about with great precision and scattering flowers among the audience as they went. At the end of the performance, they lay down on great couches, stretched out their trunks, and feasted on a banquet of fruit and vegetables, which had been laid out for them on an ornamental table in ivory and cedar. According to Pliny's *Natural History*, the Roman elephants were sometimes taught to dance or imitate gladiators in the arena. A Roman impresario once staged a comedy in which four elephants carried a stretcher on which lay a fifth, representing an expectant mother in childbed. What is somewhat harder to believe in Pliny's narrative is that some of the most skilled elephants were actually trained to walk the tightrope. Some historians have accepted this as a fact, and there have actually been rumors that a late-nineteenth-century circus elephant could perform this feat—on a very thick rope, to be sure! It would seem more prudent, in spite of the elephant's well-developed sense of balance, to await the display of a similar feat from a present-day circus elephant. Nor does it enhance the credibility of Pliny's account that he claims, in another part of his narrative, that elephants can also be taught to climb up and down ropes, something that is manifestly impossible.

After the fall of the Roman Empire, the elephant disappeared from the Western world, some early medieval writers put it on par with the basilisk, the mantichora, the sphinx, and other creatures of myth and legend. The two things every medieval bestiary reader knew about elephants were that they had no joints in their legs, which enabled hunters to capture them by cutting down the trees they were leaning toward while they slept, and that the elephants were often killed by mice crawling into their trunks, which communicated directly with the brain. This strange idea of truncal anatomy, which would put the elephant in immediate danger of literally blowing its brains out when sneezing or trumpeting violently, has remained widespread in folklore well into modern times, and the elephant and mouse motif is not infrequently met with in literature and poetry. In the year 797, Charlemagne received an elephant as a present from the Caliph of Bagdhad. It crossed the Alps with its new owner and accompanied him on several of his later travels, until its death in 810. One of its tusks was made into an enormous hunting horn; this famous "oliphant de Charlemagne" is still kept, at Aix-la-Chappelle. In 1254, an elephant from Palestine was given to King Henry III of England by St. Louis of France. It was lodged in a 40 × 20 foot elephant house near the Thames and immediately became one of the sights of London. A remarkable drawing of this elephant, by Matthew Paris, in the *Book of Additions* to his *Greater Chronicle*, is kept on permanent display at the British Library in London. In 1514, the Portuguese Ambassador Tristan da Cunha brought an elephant to Pope Leo X's menagerie at the Vatican. When the rare beast was turned over to the pope, under great festivities, the elephant bowed three times at its keeper's command, greatly impressing the church dignitaries, who praised the creature's piety. The next moment, however, the elephant filled its trunk with water from a nearby trough and soused the assembled bishops and cardinals, not even sparing the majestic figure of the pope himself.

During the 1600s, elephants became a more common spectacle in European cities. In Paris, the Versailles menagerie always kept one or two elephants. The most popular of them was an African elephant, which resided there from 1665 to 1681. With age, it became something of a gourmet, dipping its bread in huge pailfuls of soup with great relish and insisting on a ration of at least 12 pints of wine daily. In 1679, Robert Hooke recorded in his diary that he had seen an elephant in London that could "wave colours, shoot a gun, bend and kneel, carry a castle and a man etc." Another elephant was on show in Dublin. When it was accidentally burned to death in 1681, the body was used for one of the earliest studies of elephant anatomy. The popular interest for this elephant dissection was immense, and the proprietor had to employ a file of musketeers to guard the carcass until the skeleton was ready for inspection. The death of another elephant, while on exhibition in London in 1720, was attributed to the immense quantity of ale continually given to it by the spectators. No elephants reached the United States until 1796, when an enterprising sea captain named Crowninshield imported an Indian elephant whose name is unknown; there are several records of its appearance in various American towns.

The Exeter Change Menagerie

In the late eighteenth century, London had close to a million inhabitants, many of them with ample money to spend on amusements and an insatiable thirst for animal curiosities. An edict from 1697 had prohibited the exhibition of wild beasts in the streets of London, on the ground that this was the privilege of the keeper of his Majesty's lions in the Tower. But this proclamation did not prevent showmen and monster-mongers from showing their beasts at fairs and markets, or clandestinely displaying them in back alleys throughout the metropolis. It was not until 1793 that any privately owned full-scale menagerie was established in London. The itinerant showman Gilbert Pidcock had purchased a large four-story building at the Strand called Exeter Exchange (or Change). It was called this because it had been built with material salvaged from the old Exeter House, which had stood at the same site in the Strand in the late 1600s. It had once been some kind of "exchange" or station for stage coaches. The house was of considerable size, and it is likely that Pidcock first wanted it as winter quarters for his animals in between tours, but he soon realized that the animals could be profitably

The entrance to the Exeter Change menagerie.

exhibited there all the year round. Gilbert Pidcock brought with him a rhinoceros, a zebra, a kangaroo, a lynx, and a collection of rare birds; some years later, several tigers and elephants were purchased. The animals were all kept indoors, the majority of them on the first floor of the old house. The ground floor housed several small shops that flanked an arcade incorporating the Strand footway; the shop owners must have become somewhat worried when the rhinoceros shifted its weight upstairs. The cages and dens of the animals had been put in the ample parlors of the old house, whose walls had been decorated with murals of tropical motives—a sad reminder, for the wretched caged animals, of their faraway proper homes!

The old Tower menagerie went into a decline in the late eighteenth century, and Pidcock's Royal Menagerie at Exeter Change usurped its position among the Londoners. Using ostentatious handbills and newspaper advertisements to attract public attention, it became extraordinarily popular, and was a lucrative business for many years. The profits were cleverly invested by Pidcock and his deputy Edward Cross and were used to purchase a great variety of beasts, through contacts with sailors and animal dealers, that filled every apartment at Exeter Change with interesting animals and birds. It was an audacious, not to say unique, enterprise to establish a complete zoological garden in one of central London's most densely populated areas. Rather surprisingly, the civic authorities did not object to it, but there was a steady barrage of threatening letters and whining newspaper correspondence complaining about the disturbing jungle noises and noxious smells emanating from the old house. The early animal protectionists were also critical, rightly claiming that the cages and dens used at Exeter Change were far too small. Both Pidcock and Cross were skilled animal keepers, however. In a time not noted for humanity toward captive animals, they took good care of their beasts, some of which resided at Exeter Change for decades and became well-known London favorites. Several lion and tiger cubs were successfully reared there. The exhibition catalogue of the Exeter Change menagerie was quite well written, containing lengthy quotations from the natural history works of Buffon and Oliver Goldsmith as well as a dedicatory poem to the inhabitants of the menagerie:

> *And if you would wish for an exquisite treat,*
> *At nine in the Evening the Wild Beasts all eat;*
> *Their dishes so various, substantial and good,*
> *That pleasure they give while enjoying their food,*
> *And wonder impress, both delightful and strange,*
> *On each one that visits famed* Exeter Change.

In 1810, Gilbert Pidcock died, and another veteran menagerist, Stephen Polito, bought the whole establishment at auction.

When Edward Cross bought out old Mr. Polito in 1814 and became the sole owner of the menagerie, one of his ostentatious newspaper advertisements likened him to "that primeval collector of natural curiosities, Old Noah." And the boastful menagerist was not far off the mark: the old house at Exeter Change contained a remarkable collection of animals, far outclassing any other British menagerie. The huge lion Nero was a long-term favorite among the Londoners, and the exhibition contained four other African lions and several tigers, leopards, jaguars, and hyenas, as well as a boa constrictor, an orangutan, and "the greatest variety of Crocodiles ever exhibited." Antelopes, camels, llamas, bison, and sea lions were also kept. The huge collection of birds featured several African ostriches and five different species of eagles and vultures. Edward Cross corresponded with many of the leading anatomists and zoologists of London, such as Sir Everard Home, Sir Astley Cooper, and Joshua Brookes. He was also a patron of the arts, and several times allowed Sir Edwin Landseer to portray his beasts in various settings.

An Elephant in the Theatre

During the 1811 season, it was very fashionable to use animals in dramatic performances on the London stage: not infrequently, a well-trained dog, ape, or horse received more accolades than the human actors and actresses. Mr. Henry Harris, manager of the Covent Garden theater, was an audacious theatrical director who was one of the pioneers of English hippodrama: the use of horses in theatrical productions. Together with his partner, Mr. John Philip Kemble, Mr. Harris had hired horses and riders from Astley's circus to act in melodramas written particularly to suit the display of these equine performers. Elderly purists, meanwhile, were aghast at these novel vulgarities, which had usurped the place of Shakespeare's dramas at this venerable theatre. They suggested that the horses themselves should be poisoned before they poisoned the national taste, and described how Shakespeare's statue had groaned and shaken its head when being forced to endure Mr. Harris's prolonged equine histrionics. Several other establishments had followed suit, and hired various animal actors.

In late 1811, Mr. Harris, in an attempt to outdo the competition once and for all, recruited an elephant and its keeper to perform in a pantomime version of *Harlequin and Columbine.* This elephant, a young Indian male, was called Chunee; it had probably grown up in captivity in the Indian countryside. When just 2 or 3 years old, it was purchased by Captain Hay, of the East Indiaman Lady Astell. In July 1810, he took it from Bombay to London. As soon as Chunee had disembarked, he was purchased, at the docks, by Messrs. Davis and Parker, of Astley's circus, which was at this time London's leading establishment of its kind. At the circus, Chunee was just another elephant until he was rented, for the huge sum of 900 guineas, by the Covent Garden theater. Mr. Harris advertised his newly recruited elephantine actor widely, and most of the people at the sold-out premiere of *Harlequin and Columbine* had come to see the elephant. There was a general hush when the huge beast first set foot on the stage. The setting was that the sultan of Cashmere was returning from a day's tiger hunting, seated in a howdah on the elephant's back. Just before the sultan in the procession was a slave carrying a bowl of strong rum, from which Chunee helped himself to a few mouthfuls, using the trunk. The Indian elephant keeper from Astley's circus, who had joined Chunee in the elephant's theatrical career, was seated on the neck, but the audience noticed that the swarthy, turbaned individual himself looked quite frightened, and that he was holding on for dear life as the elephant hastily walked toward the middle of the stage. Chunee was ordered to kneel, to enable the sultan to dismount, and the elephant obeyed the command. The audience cheered wildly, and the clamorous sounds from galleries, pit, and boxes induced a severe attack of stage fright: the elephant abruptly jumped to its feet and pushed its way out of the stage, driving the actors and guards before it. The terrified sultan and the shrilly screaming elephant keeper were desperately clinging to their seats, and only luck prevented them from being seriously hurt.

This dismal fiasco was much laughed at in the newspapers, but Mr. Harris and his colleagues were undaunted: they kept the elephant in the cast, refusing to admit defeat and hoping that Chunee would, with time, get used to being on stage before a noisy crowd of people. Chunee's uncertain temper and lack of stage experience was a continual worry to them, however, and there were many other comic incidents, some of which would have broken the spirit of most theatrical managers. The worst thing was that the flatulent elephant's nerves suffered from the shouts and clamoring of the audience: its thunderous farts resounded in the theater hall, and there were frequent calls of "Shame!" and "Off! Off!" from the pits. The Indian elephant keeper was a brutal rascal who punished Chunee with his sharp iron goad when the elephant was disobedient. Chunee returned his dislike fully: whereas the elephant was docility itself when ridden by the theater's leading lady, the beautiful Mrs. Henry Johnston, the Indian was frequently thrown from his seat or given a resounding box on the ear from Chunee's powerful trunk. These burlesque additions to the pantomime's subject matter were greeted with cheers, applause and catcalls by the rowdy audience, which jarred Chunee's sensitive nerves further. After the play had run of 40 nights, Chunee retired permanently from the stage; the manager and actors probably breathed a sigh of relief to get rid of the unruly beast. The director of Astley's circus was not particularly interested in using Chunee in his own shows, and Mr. Norman, the circus clown, was instructed to sell the elephant. The business-minded clown struck a deal with the Exeter Change menagerie, where Chunee took up residence in a purposely built elephant den on the first floor. It was not recorded how Mr. Polito and Edward Cross managed, some time in 1812, to get Chunee up the groaning, decayed wooden stairways of the old house.

London's Favorite Elephant

From 1810 to 1826, the Exeter Change menagerie was very popular indeed, and an extremely lucrative business. This was largely due to Chunee the elephant: along with the lion Nero, he was by far the most famous inhabitant of Exeter Change. Mr. Cross's Royal Grand National Menagerie was open between 9 A.M. and 9 P.M.; it cost a shilling to see each of the three main apartments, or 2 shillings to see them all. The facade toward the Strand was decorated with colorful drawings of animal, and a large sign: "Edward Cross Dealer in Foreign Birds and Beasts." A gigantic doorman, dressed as one of Henry VIII's Yeomen of the Guard, stood at the entrance, handing out descriptive bills to all passersby and describing the living wonders awaiting them inside in a stentorian voice. Above him, a similarly noisy and gaudily colored macaw sat perched on a swing. The animals within supplied their own exotic sound effects, sometimes frightening passing horses in the Strand and making them bolt.

The animals were all fed at 8:00 P.M., at which time the whole place was crowded with spectators. The journalist William Clarke, editor of the *Cigar* magazine, wrote in his *Every Night Book* that the Lords of Parliament and the lions of Exeter Change all dined at about 8:00; the journalists reporting from Parliament thus had ample time to visit the menagerie on the way home. Himself, he vastly preferred to see a beautiful Bengal tigress and a noble

The interior of the Exeter Change menagerie; a drawing from one of Edward Cross's catalogues. Reproduced with permission of the trustees of the Victoria & Albert Museum, V & A Picture Library.

African lion gruffly debating over a bone than to hear a noble lord and a royal duke doing the same over some boring political question. Chunee the elephant rang a large bell, hanging from the roof of the den, as a signal that food was on its way. On November 14, 1813, Lord Byron visited the menagerie to watch the lions, tigers, and panthers "growl over their grub." In his diary, he recorded that the sloth much reminded him of his own valet, both in appearance and in habits. Lord Byron was much amused to see the elderly Exeter Change hyena Billy's great affection for its keeper, but the sight of the camel made him pine for Asia Minor. He also wrote that the face of the hippopotamus looked very like that of the Prime Minister, Lord Liverpool. This is a remarkable observation, since there is no record of the Exeter Change having a hippopotamus at this time. The menagerie had a rhinoceros, however, and it may be that Lord Byron's poetic imagination got the better of his knowledge in zoology; the existing portraits of Lord Liverpool give no hints either way. Lord Byron also saw Chunee perform. The keeper Alfred Copps had taught the unruly elephant a bag of tricks, and Chunee gently returned a coin Lord Byron had tossed into the den. Chunee then removed his tall hat and gingerly put it back on his head. The elephant was on its best behavior throughout his visit, and Lord Byron joked that he would like to employ it as his butler.

Chunee had an impressive memory and always remembered some person that had shown him kindness on previous visits. During the elephant's tempestous theatrical career, Chunee had made friends with several of the actors. One of them, Charles Mayne Young, had a life ticket to the menagerie, and when he visited the elephant, while taking a stroll on the Strand, Chunee always recognized him. The great tragedian Edward Kean was another of Chunee's old friends. Once, when returning from a lengthy tour to the United States, he doubted whether the elephant would remember him. But as soon as Chunee heard his voice, the elephant turned around in amazement and welcomed him with many caresses. Another London man-about-town, Thomas Hood, the poet of the Learned Pig, was also an avid visitor to the menagerie and a particular friend of the Chunee:

I was the Damon of the gentle giant,
And oft has been,
like Mr. Kean,
Tenderly fondled by his trunk compliant.
Whenever I approached, the kindly brute
Flapped his prodigious ears, and bent his knees . . .
I bribed him by a apple, and beguiled

The beast of his affection like a child;
And well he loved me till his life was done
(Except when he was wild).
It makes me blush for human friends, but none
I have so truly kept or cheaply won!

Edward Cross was a clever, observant man, and although he lacked a formal education, his knowledge in practical zoology was considerable. He was known and respected by many of London's anatomists and zoologists and was a particular friend of the surgeon Sir Everard Home, who was a copious writer on comparative anatomy and physiology. Home was often consulted when men or beasts belonging to the menagerie were taken ill; in return, many dead animals of comparative anatomical interest were taken to Home's headquarters at the Royal College of Surgeons in Lincoln's Inn Fields to be dissected and prepared for the famous Hunterian Museum. Sir Everard was something of an elephant enthusiast, and he visited Chunee many times. He was particularly interested in the elephant's power of digestion. In 1823, Sir Everard wanted to investigate whether the elephant was capable of appreciating music. Mr. Cross was easily persuaded to assist him, and he recruited a pianoforte tuner to bring his instrument to Chunee's den. Sir Everard accompanied him on the French horn when they started playing, and Chunee seemed to appreciate their impromptu concert: "He brought his broad ears forward, remained evidently listening, and he made use of sounds rather expressive of satisfaction than otherwise." For a lark, they also played for the lion Nero, which stood glaring at them when the high notes were sounded. As soon as the flat notes were played, the lion sprang up, lashed his tail, roared fiercely, and endeavored to break free, to the alarm of the female spectators and the amusement of Sir Everard Home.

Big Trouble at Exeter Change

Each day, Chunee consumed more than 800 pounds of hay, corn, straw, carrots, mangel wurzel, and biscuit. He grew at an astounding rate and in 1820, when a new den had to be built to accomodate the huge beast, had more than doubled his size in a little more than 8 years. When Mr. Cross published a new guidebook to Exeter Change in 1820, he proudly declared that Chunee was Europe's largest elephant, being more than 10 feet in height and weighing about 5 tons: "familiarly speaking, he may be called an animated mountain." In spite of his huge size, the elephant was timid and easily frightened, Mr. Cross claimed. Once, when quite young, the elephant had been terrified by a large dog leaping forward; now, it was enough that Mr. Cross's fierce little fox terrier ran into the den for Chunee to cower into a corner, trumpeting with fear. In his guidebook, Mr. Cross advised the visitors not to try to pat or fondle the elephant; he had received many complaints from ladies whose white gloves had been ruined by the smelly oil rubbed into Chunee's coarse hide by the keepers. Other well-intentioned advice was that the ladies should not stand too close to the cage of Ould Bill, the orangutan, since he liked to dash forth and grab hold of their long skirts. It was also strongly recommended to get out of the way sharply when the llamas cleared their throats to spit.

In his pamphlet, Mr. Cross boasted of the elephant's docility and its great affection to its keeper, but he neglected to mention that there had been several violent incidents at Exeter Change, all involving the elephant. Chunee's first keeper Alfred Copps long seemed capable of controlling his charge; several visitors to the menagerie had, like Lord Byron, even been quite impressed by the elephant's obedience and sagacity. But one day in early 1815, Chunee attacked Copps without warning and gored him into a corner of the den with great force; it was fortunate that the tusks went on each side of his body. Copps fell senseless to the floor, and Chunee proceeded to throttle him using the trunk. A bystander called Mr. Cross, who valiantly ran into the den and struck Chunee with a pitchfork. The elephant turned and attacked him instead, and the senseless Copps, who had, according to another keeper, "scarce any breath remaining in his body," was pulled out through the bars. Rather understandably, Alfred Copps did not dare to enter Chunee's den again, and he left Exeter Change for good, later to become head keeper of the Tower menagerie. A young man named George Dyer was appointed in his place. This individual had little previous experience of animal keeping, and had probably never seen an elephant before; he completely lacked the skill to control the unruly beast, and Chunee's behavior went from bad to worse. The gentlemen who, like Lord Byron, let Chunee act the part of their valet, sometimes received their hats back in a flattened and dirty condition, or covered in elephant dung. When Dyer objected to this latter outrage, and struck Chunee with a poker, he was soused with dirty water that the elephant had drawn into its trunk. The wretched man had his own bedroom next to the elephant's den, which is unlikely to have benefited his night's sleep. Chunee could easily reach into his room with the trunk, and once tore up the keeper's complete wardrobe. George Dyer complained to Mr. Cross, who, rather surprisingly, did not fire this bungling youth; instead, an experienced animal keeper named John Taylor was employed as his assistant. According to a newspaper clipping, Mr. Taylor was a former circus artist who had really dedicated his life to animal keeping; although a lion had once torn off his right arm, and eaten it before his eyes, he had not despaired of the inherent docility of the brute creation. The one-armed old keeper was no fool, however, and it was apparent to him that Chunee had been ill treated by his brutal and inexperienced attendant. By adopting a policy of kindness and constantly staying with the elephant, Taylor had considerable success, and Chunee's behaviour steadily improved. After the Battle of Waterloo, the patriotic Mr. Cross invited a large force of soldiers and Bow Street constables for a free showing of the menagerie. One of the policemen was unwise enough to tease the elephant with grimaces and threatening gestures. Chunee suddenly grasped him with the trunk and tried to pull him through the bars of the den; if old Taylor had not been quick to intervene, the man might have had to pay a high price for his jokes.

One evening, George Dyer attended a masquerade at the English Opera House. He returned early in the morning, having eaten and drunk well. He was dressed in an extravagant devil's costume and imprudently went into the elephant house to reach his own quarters behind it. The startled elephant did not recognize him, and Dyer received a violent blow from the trunk, which disfigured him for life with a pugilist's nose. The enraged Dyer punished the elephant severely, and Chunee took an even more determined dislike to him, frequently striking at him with the trunk or squirting him with water and refusing to take any food from him. George Dyer again complained to Mr. Cross about the many hardships of his hazardous employment, but Taylor pointed out his shortcomings to Mr. Cross, and Dyer was unceremoniously discharged. In his place, John Taylor was promoted to head elephant keeper, with a German youth named Johann Tietjen as his assistant. Old Taylor was skilled in handling all kinds of animals, and his reign as elephant keeper would have been a happy and peaceful one, had his unexpected promotion to head keeper not given him a taste for office intrigue. He seems to have been a small-minded, gossipy character who was not above calumniating his colleagues to gain favor with Mr. Cross. Later in the same year, he gained further promotion and became deputy to Mr. Cross, but the other keepers had had enough of him. They threatened to resign if Taylor was kept on the establishment, and after a tempestous meeting, Mr. Cross sacked the one-armed old man, on the spot. A man named Richard Carter (or Cartmell) was appointed his successor as elephant keeper. In his autobiography, John Taylor spared no invective in describing the sloth, wrong-headedness and cruelty of Mr. Carter, whose only previous merit was a very short tenure at the Tower menagerie. Other sources agree that Carter was a coarse, brutal character and that he could not manage the elephant in any degree. When Taylor visited Chunee 4 months after being fired from his position, he was pained to see that Carter was threatening the elephant with a sharp spear to make it perform its usual tricks. When Chunee confidingly stretched out his trunk to Taylor, the keeper saw that it had a lump the size of an egg from a thrust of the spear. Taylor expostulated with Carter and Mr. Cross but received gruff replies. Later, he wrote to William Tyler, a veteran animal keeper who was Mr. Cross's deputy, pointing out Carter's incompetence and demanding to be reinstated as elephant keeper in his place, but his lengthy missive was not even answered. Carter's ill treatment of the elephant led to further trouble at the menagerie, particularly during the musth, the elephant's annual period of heat. In India, the elephants were sometimes let loose in the forests during this period, but it was of course impossible for Mr. Cross to let the ungovernable elephant roam the streets of London at will. Richard Carter and Mr. Cross had a theory that libido and constipation could be treated by the same means. They mixed large amounts of calomel, Epsom salts, syrup, and croton oil into poor Chunee's molasses; it was recorded that this dose of laxatives would have purged some six thousand human beings. The treatment had the desired effect, and one would not envy the keepers whose duty it was to clean out the den after this elephantine purgation. Carter believed that Chunee had been somewhat calmed by this unsavory physic, which was repeated every year during the rutting season. Modern zoologists would not agree with him, but the dosages used were extremely high, and might well have killed a less hardy beast; the peristaltics of the bowels continuously operating at maximum velocity might well have had a fatiguing influence even on an ungovernable elephant.

The German keeper Johann Tietjen remained as Carter's assistant. The reason for this, according to the quarrelsome John Taylor, was that Carter was quite incapable of handling the elephant himself. On November 1, 1825, Tietjen entered the elephant house early in the morning. A few spectators had already gathered outside, and the German wanted to show off his mastery of the huge beast. After the previous incidents, Mr. Cross had ordered, doubtlessly due to the insistence of Carter, that the keepers always enter

Chunee's den in pairs, one of them flourishing a 12-foot spear to keep Chunee "in awe" while the den was cleaned. Tietjen told Carter to drop the spear, however, since he trusted the elephant fully. Chunee then took the spear in his trunk and playfully whirled it round, to the terror of Carter, who was standing nearby. Being told to do so by Tietjen, Chunee obediently dropped the spear. Mr. Cross was just then passing by and he teasingly called out to Tietjen: "Don't go near him, John, or perhaps you may have your sore foot trod on!," since earlier the same morning, one of the gnus had trod painfully on poor Tietjen's toes. The German was cleaning the elephant with a broom and told Chunee to turn around. The elephant did so very sharply and thrust his tusk through Tietjen's ribs. Carter screamed and rushed out of the den, but the valiant Mr. Cross, who heard the horrid sound, rushed back to pull the lifeless keeper out through the bars. Chunee did not try to attack him or trample on Tietjen's body; indeed, the elephant stood quite still, seemingly aghast at what had occurred. A medical practitioner was called, but Tietjen was stone dead. According to the old-fashioned British laws, the elephant was summoned before the coroner's jury, as it had caused the death of a man. At the inquest, the keepers testified that the elephant had always seemed quite attached to Tietjen, and the spectators affirmed that, immediately after the fatal blow, the elephant had hastily withdrawn to a corner, trembling all over. The verdict was death by accident, and the jury laid a nominal deodand (or fine) of 1 shilling on the elephant. There was much writing in the newspapers about this "melancholy occurrence." The old elephant keeper John Taylor put the blame on Carter's ill treatment of the elephant, but Carter himself (probably erroneously) suspected that Chunee had not recognized Tietjen, since the accident-prone German keeper had spent the previous 3 weeks in bed, having been badly torn up by one of the leopards.

The Destruction of Chunee

Throughout the early 1820s, Chunee kept his position as London's favorite animal. A whole generation of English children were taken to see the elephant, Princess (later Queen) Victoria, Charles Dickens, and Robert Browning among them. Another early visit to Chunee was described in Douglas Jerrold's *Punch's Letters to His Son:* a young boy threw a halfpenny coin to the elephant, and Chunee seized it with an elegant curve to the trunk to give it to the cake seller outside in exchange for a small biscuit, which was at once put into his capacious mouth. In the hot summer evenings, many strollers on the Strand stopped by to see their old friend the elephant, but even more of them went on to the English Opera House nearby, where the popular actor Charles Mathews held his popular "at Home" entertainments. In July 1825, Chunee's particular friend Thomas Hood published another comic poem in the *London Magazine,* in which the elephant, envious of his rival's fame, suggests to Mr. Mathews that they should cooperate in the same show instead of competing.

Mr. Cross was severely shaken by the death of poor Tietjen. He swore that the elephant would one day be the death of him and declared that he would have it destroyed before it committed any more mischief. When a visiting American offered him 500 pounds for the elephant, Taylor suspected that Mr. Cross would have accepted the offer, but the negotiations were stranded owing to difficulties in getting Chunee downstairs and the reluctance of any American sea captain to transport a 5-ton elephant across the Atlantic. Mr. Cross's worst nightmare was that the elephant would break free in central London, causing death and carnage in the metropolis. In Venice, 6 years before Tietjen's death, an elephant belonging to a Swiss lady, Mlle. Garnier, had killed its keeper and held the city under a rule of terror for more than a day. Chunee's old friend Lord Byron was just then in Venice for a visit, and he amused himself by watching, from the safe distance of his gondola, the enraged beast flinging great beams of wood into the water. When the elephant charged, the Venetian soldiers dropped their muskets and ran. The conclusion was that, in the rather flippant words of Lord Byron, "the elephant dispersed all his assailants and was at last killed by a shot in his *posteriore* from a field-piece brought from the *Arse*-nal on purpose." Another elephant, under the management—or rather lack of management, it would seem—of the same lady broke free in Geneva the year after. It roamed the streets at will and, after attempts at poisoning had proved futile, was also killed by a well-aimed cannon shot. Some Swiss gentlemen urged Mlle. Garnier to devote her energies to matters more suitable for the female mind than elephant keeping, before her gunned-down elephants would clutter the streets of every European capital.

Mr. Cross was well aware of the risk of the elephant breaking loose, but he believed that the huge beast was secure up in his den on the first floor. In spite of the death of Tietjen, Chunee kept his popularity among the Londoners, and Edward Cross was loath to have the leading light of the menagerie destroyed or removed. On Sunday, February 20, 1826, Chunee again showed signs of anger and irritation and gored the walls of the den until the floor was covered with plaster and mortar. Richard Carter tried his old trick of mixing a strong dose of laxatives into the elephant's gruel, but this time the medication seemed wholly ineffective. To speed up the actions of this "opening medicine," Carter offered Chunee a large tub of strong, hot ale. The elephant had grown suspicious of his keepers, and it was not until Carter himself had drunk a jug of the ale that Chunee emptied the tub, with audible satisfaction. The ale seemed to have a calming influence on the elephant, but on Wednesday, Chunee was more furious than ever. The den had been constructed by the master builder Mr. Harrison, who had ensured Mr. Cross that it would withstand the assault of any animal without even being dented. It was unfortunate for Mr. Cross that this guarantee had not been put in writing, since on Wednesday morning, the elephant again battered the sides of the den, cracking a thick wooden beam and pulling down parts of the roof. Mr. Cross and the keepers feared the worst when they saw the infuriated state of the elephant. If the great beast broke free and shook the whole building down, several lions, tigers, crocodiles, and boa constrictors would be liberated in the process, to wreak havoc among the Londoners. After a brief council of war, they decided to kill Chunee by poison. Carter the elephant keeper vigorously seconded this motion, but poor Mrs. Cross, who was very fond of the elephant, wept bitterly.

At Wednesday noon, Edward Cross went to the pharmacy to purchase all potent poisons in the premises; the startled apothecary first refused to sell him anything, believing him to be an intended mass murderer. After finally making the purchase, the keepers mixed corrosive sublimate into the elephant's hay, but Chunee refused it; nor did the sagacious animal touch a trough full of arsenic and gruel. Mr. Cross then tried another ruse: he persuaded a boy, who was not connected in any way with the menagerie, to pose as a visitor and to give the elephant several large buns, one of which was filled with poison. Chunee eagerly took each of the buns until handed the poisoned one; the wary beast dropped it and trod it underfoot and afterward took no food from anyone. Some hours later, Chunee again attacked the wall of the den, cracking another wooden beam and weakening some of the others; bricks and mortar rained from the roof and the whole building shook, as if hit by an earthquake. Pandemonium broke out at Exeter Change: the few remaining spectators were hastily evacuated, and the keepers armed themselves with cutlasses, bayonets, and harpoons. Mr. Cross ordered his deputy, William Tyler, to secure the sagging walls of the elephant house with some strong rope and to lash down the gate of the den. Mr. Cross himself dashed off to call in the police and the military. Only the part of the floor directly under Chunee's den had been reinforced to support the elephant's weight, and the rest of the decayed flooring would immediately collapse under Chunee's weight. This would cause the elephant to fall into the arcade below, along with many of the other animals residing on the first floor; their dens would present but feeble barriers to Chunee's great strength. Outside the gates to Exeter Change, an eager mob was gathering, alerted by the elephant's trumpeting and the deafening roars of the lions and tigers. They were speculating that the whole place would be shaken down and that the entire company of animals would come bursting out through the great gate into the Strand, the elephant leading the assault.

Mr. Cross first ran to the Paddington police station, where his brother-in-law Mr. Herring and two other members of the Bow Street patrol seized their truncheons and marched towards Exeter Change. When they arrived, the whole place was in confusion—people milling about and the animals most uproarious. According to an eyewitness, "the combined roaring of the Lion, Elephant, Tigers, and other enraged animals was dreadful." The keepers, led on by Tyler and Carter, desperately tried to reinforce the broken bars of Chunee's den with rope, while the elephant's furious assault shook the foundations of the old building. Mr. Herring now understood that the situation was a precarious one and that their truncheons were useless against such a formidable opponent. Some rifles were fetched, and the three policemen advanced on the elephant while the vile Carter called out "Chunee, Chunee, Chuneelah!" to make the elephant present a favorable target to them. The policemen fired their rifles, expecting the elephant to drop dead on the spot, but Chunee instead gave a furious hiss, dashed after them and struck at them with the trunk, missing Mr. Herring by a narrow margin.

In the meantime, Mr. Cross was visiting his old friend, the anatomist Joshua Brookes, who was at that moment delivering a lecture to more than a hundred medical students at his Marlborough Street anatomy school. When the pale-faced, distraught menagerist ran into the theater, calling out "Sir, a word with you, if you please, immediately: I have not an instant to lose!," every one present knew that some momentous happening was afoot. Brookes and his students listened with rapt attention while Mr. Cross explained his

Destruction of the Noble Elephant, one of the engravings of Chunee's death. Reproduced with permission of the trustees of the Victoria & Albert Museum, V & A Picture Library.

extraordinary predicament. Brookes went with Mr. Cross to advise the policemen how to aim for the balls to hit vital inner organs. The students ran through London in the direction of Exeter Change, whooping and carousing to celebrate their unexpected half holiday and spreading the news of the berserk elephant to all passersby. Mr. Cross had also called in a troop of soldiers from the Somerset House military depot, and they elbowed their way through the crowd outside, carrying their muskets. After the civilian firing squad had failed in its mission, the soldiers knelt before the den and discharged an uneven volley, which was barely noticed by the elephant. These soldiers were no hardened veterans from the Battle of Waterloo but rather a troop of raw recruits with hardly any training; many of their shots actually missed the elephant, which dashed about the den "with the speed of a racehorse, uttering frightful yells and screams, and stopping at intervals to bound from the back against the front." These bungling recruits had only three cartridges each, and they rammed such generous amounts of gunpowder into their muskets that the barrels would have burst had the policemen not intervened and loaded their weapons for them.

Mr. Herring sent for all the powerful rifles and muskets in Mr. Stevens's gunsmith's shop in High Holborn, which were distributed among the policemen, soldiers, keepers, and other individuals who had courage enough to remain in the room. In all, fourteen people were armed. Outside, the crowd was held back by two other patrols of Bow Street officers, who were offered bribes of 1, or even 2, guineas, by rich snobs who wanted to see the elephant being killed. When Mr. Herring and his men had fired upward of fifty rounds, Chunee suddenly fell to his knees with a groan. Herring dashed forth, brandishing a harpoon, calling out "He's down, boys! He's down!" He tried to climb through the broken bars of the gate but had difficulties getting into the den; this turned out to be very fortunate for him, since Chunee rapidly leapt up again and made another furious assault on the gate. The remaining bars were broken and the entire door lifted off its hinges with great force; Mr. Herring, who had not been able to release his hold of it, was catapulted out into the room like a rag doll. Had the door not been secured with powerful ropes and chains, nothing would have prevented Chunee from bursting out into the room.

The riflemen kept up their barrage of fire, and the keepers tried to thrust spears and harpoons into the elephant's flanks, but with little effect on the

Another, less skilled anonymous engraving of the elephant's last struggle for freedom. Reproduced by permission of the British Library.

An engraving from George Cruikshank's drawing of Chunee's destruction. From the author's collection.

furious elephant's onslaught. Being aware of the means of destroying Mlle. Garnier's beasts in Venice and Geneva, the distraught Mr. Cross dashed off to a nearby military depot to fetch a cannon. Chunee turned about at great speed, and the riflemen were unable to take a steady aim for the windpipe, eye, and ear, according to Joshua Brookes's instructions. Carter, the dastardly elephant keeper, thought of a cunning plan. Using the same commands as during the shows, he ordered Chunee to kneel. Nobody expected the elephant to obey, but Chunee really knelt. Carter then thrust a long bayonet into Chunee's flank, and Herring fired his powerful rifle at the ear at short range. With a furious roar, Chunee thrust his weight against the front of the den in a final, desperate bid for freedom; the door was once more lifted from its hinges and was held only by the ropes and chains securing it to the gates. Chunee reeled into the center of the den and slowly sank to the floor, motionless, although his puny assailants fired their rifles at short range and thrust harpoons and spears into the elephant's unprotected flank. When Mr. Cross returned a few minutes later, with a cannon and artillerists in tow, the pride of Exeter Change had already breathed his last. During the melee, not fewer than 152 projectiles had been fired.

Chunee Is Used Up

Poor Mr. Cross was seized with despair when he saw the horrid spectacle: the lifeless elephant lying in his blood-spattered den, and the exulting killers, Herring and Carter, congratulating themselves that the furious beast did not break free. This dismal scene was accompanied by the thunderous roaring of the other animals and the shouts of the populace outside. Chunee lay fallen like a mountain, his sinewy proboscis remisely outstretched among the blood and gore. Around the great carcass, the triumphant human mice were cavorting, yet the names of these elephant killers lay hidden in the faded pages of old newspapers. The name of Chunee, however, became a household word in London for many years.

Only with the greatest difficulty could Tyler and Carter persuade Edward Cross to open the big gates of the menagerie and let in the noisy crowd waiting outside in the Strand. Cheering wildly, the rowdy mob broke past the policemen and ran upstairs. They stopped short, however, in spite of themselves, at the awe-inspiring sight of the fallen giant in the blood-spattered, broken den. Although they were charged the ordinary entrance fee, more people visited the menagerie that evening than during an entire normal week. Owing to the objections of the crowd, Mr. Cross could not close the doors until midnight. In the following days, the dead elephant had as many visitors. The wretched Mr. Cross, who was pondering how best to dispose of the 5 ton cadaver, was somewhat consoled by the cash flowing in. Sir Humphry Davy, President of the Royal Society, the legislator Lord Stowell, and the Bishop of London all honored the elephant's lying-in state with their presence. Chunee's old friends from the theater grieved him bitterly, as did Thomas Hood, who seems to have visited Exeter Change shortly after the elephant was killed. He was inspired to write the following poem:

The very beasts lament the change like me.
The shaggy Bison
Leaneth his head dejected on his knee;
The Hyæna's laugh is hushed; the Monkeys pout;
The Wild Cat frets in a complaining whine;
The panther paces restlessly about,
To walk her sorrow out;
The lions in a deeper bass repine;
The Kangaroo wrings its sorry short forepaws;

Shrieks come from the Macaws;
The old bald Vulture shakes his naked head,
And pineth for the dead;
The Boa writhes into a double knot;
The Keeper groans
Whilst sawing bones
And looks askance at the deserted spot . . .

Less sentimental individuals made offers to purchase parcels of elephant meat, and the *Mirror* newspaper published a series of recipes for elephant steaks and stew. Every medical student in London wanted to be present at the dissection of the elephant, and the shrewd Joshua Brookes charged them a considerable fee for this privilege. At this time, phrenology, a pseudoscience claiming that the individual's character was determined by the shape and irregularities of the skull, was in its heyday. Several practicing phrenologists were active in London. One of them, a certain Mr. Deville, apparently considered veterinary phrenology as one of his subspecialities, since he wanted to try his art on the elephant's head. After a huge craniometer was applied to the prostrate Chunee's head, Mr. Deville wanted to take a mold of it, to produce a cast of the elephant's skull for his private museum. The mold was taken from the head to the shoulders, and 7½ hundredweight of plaster of paris was required. Mr. Deville's conclusion was that the "bump of fury" on the elephant's head was extremely well developed. He hinted to Mr. Cross that he might—for a small fee—examine all the other animals to find out if any of them had any violent tendencies, but Mr. Cross was a down-to-earth man who distrusted this modern, unproven branch of science. Another phrenologist, the well-known Dr. Spurzheim, wanted to purchase Chunee's brain for dissection, but Mr. Cross was again unresponsive, since he wanted to keep the skull intact. Chunee's old friend Sir Everard Home, who had examined the elephant several times during life, was allowed to take several specimens of muscle tissue from the elephant for use in his experimental studies on the structure of muscular tissue of various animals.

Mr. Cross left the elephant on its bizarre *lit de parade* for several days, until the overpowering stench from the cadaver prevented all but the most hardy Londoners from paying their respects. In the meantime, the neighboring houses had been evacuated by their furious tenants, who threatened Mr. Cross with criminal prosecution. The shops on the ground floor, immediately below the putrid elephant, were deserted by their owners. The animals also suffered: it is recorded that one of the tigers residing in a room next to Chunee's den fell dangerously ill and was carried away on a stretcher. By Saturday, March 4, several blocks in central London had become uninhabitable. Sir Richard Birnie, one of the Bow Street magistrates, sent a message to Mr. Cross "that, unless the body was removed by Monday morning, Mr. Cross would hear from Sir Richard, in a way he would not like." This thinly veiled threat had the desired effect. Two huge pillars were erected on each side of the battered den to support a huge cross beam, from which a pulley capable of raising 10 tons was suspended. The body was first turned by ropes fastened to all four legs. Then, the trunk was cut off and the eyes gouged out, since these parts had been sold to gentlemen within the audience. Joshua Brookes made a deep incision through the abdominal wall, and the entire contents of the abdominal and pelvic cavities were taken out. The elephant's body was again lifted, and nine butchers worked day and night to flay the enormous cadaver. On Sunday morning, Chunee's entire hide (save for the trunk) was dragged to the residence of a certain Mr. Davis, who had purchased it for 50 guineas. The entire intestinal tract of the elephant had also disappeared by this time; there were dark rumors that Mr. Cross had had it dumped into Thames from Waterloo Bridge, under the cover of the night.

On Sunday morning, Joshua Brookes started the dissection, assisted by Dr. Spurzheim, the zoologist Mr. Ryals, and several other anatomists; more than 100 medical students were also present. The educational value of this spectacle was debatable, however, owing to the putrid state of the animal and the extreme speed at which the dissection was performed. Joshua Brookes was egged on by Edward Cross, who wanted to avoid criminal prosecution at all cost. According to an eyewitness, who wrote one of the many pamphlets about the elephant's violent demise, the sight was not a prepossessing one: "A number of medical men were hacking most furiously at the enormous carcass of the animal, cutting away pieces of flesh that weighed at least 10 or 12 pounds each. It stunk most intolerably, and was getting putrid and black." The groaning medical students carried huge burdens of meat down into the entrance to the Strand, where slaughtermen were busy cutting it into smaller pieces. According to a contemporary account, the dirty, exhausted, cursing butchers cutting into the putrid meat presented a particularly disgusting appearance, although this did not prevent them from being closely surrounded by a curious crowd. Newspaper men were shocked to detect several ladies among them, "eager to witness the pleasing *sight* and *smell*, certainly a delectable spectacle to *delicate* minds."

It had been suggested to Mr. Cross that the elephant meat be used to feed the other beasts at the menagerie, but he did not allow this, probably less due to sentimentality than to the extremely unwholesome aspect of the meat and grease. Instead, the burdens of meat were hauled off by an endless caravan of carts, to be delivered to various purveyors of cat's meat. The elephant beef Mr. Cross had considered too seasoned for his big cats at the menagerie thus had to make do for their smaller cousins on the streets of the metropolis! One has to admire his skill in finding a use, and a buyer, for even the most unappetizing parts of the beast's anatomy. The following morning, many of London's innumerable cats—approximately two hundred thousand in number some decades later, according to Henry Mayhew—were allowed to taste an unique meal, consisting of grease, lard, and putrid elephant's meat, which the half-starved felines are likely to have devoured with as much avidity as the present-day pampered pussies taste their canned "casserole of duck in gourmet sauce."

During the dissection, Joshua Brookes found nothing remarkable, except that Chunee's entire body was riddled with bullets. The elephant's heart was nearly 2 feet long and 18 inches broad. When the pericardium was opened, it was observed to contain about 5 gallons of blood, the effusion of which had been caused by one or more sharp objects penetrating the heart. Several of the anatomists and zoologists present remarked that the state of the body, in all particulars, denoted the most perfect health as could be expected from a 20-year-old Indian elephant in the prime of life. Chunee would, under normal circumstances, have had many years left to live.

What Happened to Chunee's Remains?

Although Chunee's death had deprived Mr. Cross's menagerie of its prime attraction, the Exeter Change remained popular for several years; it had become something of an institution among the Londoners. Although both the British Museum and the Royal College of Surgeons were reported to have shown some interest in purchasing Chunee's mounted skeleton, Mr. Cross eventually decided to keep it. This is likely to have been a clever move, since Chunee continued to serve him well. There was, for several months, a steady flow of visitors coming to see the skeleton standing in the battered den as a memorial of the death of the martyred elephant. Several people tried to persuade Edward Cross to get a replacement for Chunee, but he was, naturally, most reluctant to keep another elephant indoors, in view of the tumultuous and distressing scenes of 1826. Chunee's martyrdom was in fact a close harbinger of the end for the menagerie house. In 1828, the Strand was to be widened as a part of the urban improvement scheme leading to the construction of Trafalgar Square, and the Exeter Change building was marked for demolition. Mr. Cross remained convinced that it was quite possible to keep a complete indoor zoo within a central London building, and his animals were moved to another Noah's ark situated a few blocks away, at Charing Cross. His old friend Thomas Hood commented on this in another poem:

> *Let Exeter Change lament its change,*
> *Its beasts and other losses —*
> *Another place thrives by its case,*
> *Now* Charing *has two* Crosses.

On the way to Charing Cross, an antelope and a hyena escaped, but Mr. Cross and the keepers were able to recapture them after a furious chase through the London streets. The great procession of animals and the playful fugitive hyena inspired another broadside print, called *An Uproar on Change or a Trip from Exeter to Charing Cross.* This fanciful aquatint depicts the animals galloping off to their new abode, led on by the hyena and antelope. Even the

Uproar on a Change, an engraving of the fugitive animals leaving their new Noah's Ark at Charing Cross. From the author's collection.

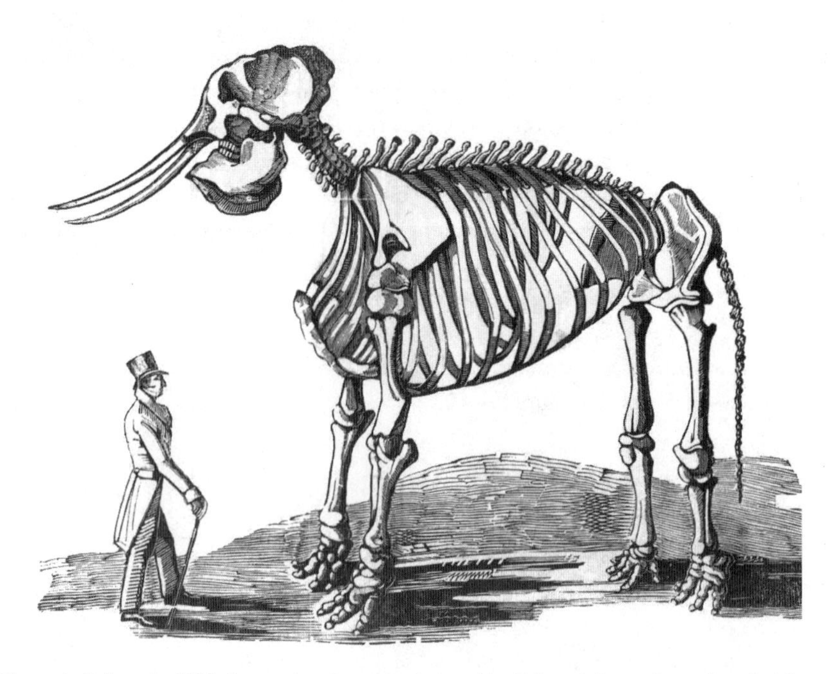

Chunee's skeleton in 1829, from a drawing commissioned by Edward Cross. Reproduced with permission of the trustees of the Victoria & Albert Museum, V & A Picture Library.

elephant skeleton is to be seen, walking out of the side door of the old Exeter Change building.

Chunee's remains did not stay long at Charing Cross. In 1829, the mounted skeleton was sold to a certain Mr. Bentley, who is likely to have been an itinerant showman, since he at once took it on tour to the provinces. Advertisement posters from Manchester and Liverpool that were kept both reproduce an engraving of the skeleton, which had probably once been made at Mr. Cross's orders, to be sold at the menagerie. An original of this engraving is in the Enthoven Collection; it shows not only the impressive proportions of the skeleton but also that Mr. Cross had apparently equipped it with two new tusks. One of Chunee's original tusks had been injured in some way when the elephant was quite young, and the other was completely broken off during the final assault on the elephant house. Mr. Bentley seems to have been more of an enthusiast than a businessman, and the tour was a fiasco: he earned very little money, and in 1830 the skeleton was lent to London University for a niggardly sum. Mr. Bentley eagerly sought another buyer, but there were few prospective purchasers capable of finding accommodation for a skeleton this size. Mr. Bentley had to repeatedly lower the prize: he himself had paid 300 pounds for it but was now prepared to accept 200 pounds. In May 1831, he contacted Mr. William Clift, the Conservator of the Hunterian Museum of the Royal College of Surgeons at Lincoln's Inn Fields. Mr. Clift recommended that the Trustees of the Hunterian Collection make the purchase, and 200 pounds were voted to Mr. Bentley, on condition that he himself saw that the skeleton was mounted in the main hall.

William Clift, who was an expert zoologist, examined Chunee's skeleton thoroughly. He was particularly interested in the short, broken original tusks, which Mr. Bentley delivered in a separate parcel. One of them showed signs of an advanced state of inflammation of the large matrix of the tusk, also engaging the jaw skeleton. This must have been exceedingly painful and is likely to have contributed to Chunee's paroxysms of fury. The eccentric naturalist Frank Buckland, who had also examined the skeleton, was even more outspoken. In his *Curiosities of Natural History,* he wrote that the barbarous murder of poor Chunee at Exeter Change remained an everlasting disgrace to the individuals involved in the slaughter: "This poor elephant was mad, but he was *mad with the toothache.*" He claimed that if an incision had been performed, down to the root, the pus would have been allowed an outlet, and the intense pain would have been greatly relieved. Frank Buckland had no suggestion for how such an incision should have been performed, before the age of anesthetics, on an elephant furiously dashing about in its den and striking at people with the trunk.

Several pieces of Chunee's anatomy, like the eyes, the trunk and the heart, were preserved during necropsy. It is unclear, however, whether they were intended, in a "pickled" condition, for a shelf of anatomical curiosities, or whether some joker wanted to serve his friends a tasty elephant's trunk and tripe pie. At any rate, no trace of either of these specimens remains today. Nor is it known what happened to the elephant's hide, which was purchased by Mr. Davis in 1826. Three years later, it was resold at an auction in London and was stated to be 30 feet in length. It was again put on the market in 1832 and fetched 32 pounds, 12 shillings, and sixpence. A note in the *Times* of October 9, 1832, states that the skin weighed 269 pounds and that it had been tanned at Greenwich. There is no trace of its further whereabouts. For many years, Chunee's skeleton remained at the Hunterian Museum. In a late-nineteenth-century drawing of this museum, reproduced in my earlier book *Cabinet of Medical Curiosities,* the skeleton is seen towering over all the other animal and human specimens, in the middle of the great exhibition hall. During the London Blitz of 1941, the Royal College of Surgeons building was struck by several German bombs, and the 115-year-old elephant skeleton was blown completely to smithereens. It is commonly believed that no fragment of it was saved: at least no exhibit related to the elephant's skeleton is present at the Hunterian Museum today. There are rumors among some British historians of zoology, however, that part of one of Chunee's tusks, perhaps the one with the marks of the fatal abscess, were salvaged by a fireman; it is reported to belong to a private collection in London.

The Immortal Chunee

One of the purposes of an early-nineteenth-century zoological garden was to demonstrate that man could shape nature after his own mind: wild beasts were kept in cages along a pleasant walk, decorated with exotic plants and shrubberies. The Exeter Change menagerie was a further travesty of nature: the animals were kept indoors in the old house's apartments, and the tropical plants were kept in pots or painted on the wall murals. The lion Nero was particularly tame and was allowed to walk at will among the visitors. Everyone was allowed to handle or feed the beasts, at their own risk. Chunee resided at Exeter Change for 14 years. His size and fame grew simultaneously, and by the 1820s he had become something of a national institution. The news of "the Destruction of the Berserk Elephant" was a shock to Chunee's many friends in London. There were several colored engravings depicting the gory details of Chunee's battle for freedom, and at least four pamphlets describing the life and death of the noble elephant. One of these engravings was available for sale only 40 hours after the death of the elephant and was heralded as a triumph for modern lithographic art.

After Chunee's tragic demise, elephant mania ruled among London journalists. All the papers, the *Times* not excluded, ran detailed reports of the elephant's last days. Tough, unsentimental reporters, who used to chronicle the latest murders and rapes of the Metropolis with gusto, now composed tear-jerking sob stories about the martyred elephant. One sentimental poem began with the words: "Farewell, poor Chuny! generous beast, farewell!" Another newspaper writer solemnly hoped that God would smite the wretched keeper who had harpooned poor Chunee after the elephant martyr had obediently knelt in front of him. In his dedicatory poem to Chunee, Thomas Hood mocked these sentimental excesses, comparing the elephant's death to another recent sensation: the hideous mass murder at Ratcliffe Highway, in which the entire Marr family had been butchered in cold blood by an unknown attacker:

When, like Mark Anthony, the keeper showed,
The elephantine scars!—
Reporters' eyes
Were of an egg-like size;
Men that never wept for murdered Marrs!
Hard-hearted editors, with iron faces
Their sluices all unclosed
And discomposed
Compositors went fretting to their cases!

Several newspaper writers criticized Edward Cross for his, in their opinion, premature decision to have the elephant shot, and some know-it-all busybodies suggested medication or laxatives that should have been tried instead. The ex-elephant keeper John Taylor wrote a lengthy pamphlet, depicting himself as Chunee's last loyal friend among a crowd of villains, fools, and incompetents and roundly accusing the later keepers of having precipitated the cataclysm through ill treating the favorite of all Londoners. Taylor was apparently still out of a job and in dire financial straits; the title page of his memoirs stated that it was published "for the benefit of the author, a person deprived of one arm." Taylor's stories were avidly picked up by the press, and both Mr. Cross and the elephant keeper Richard Carter were booed by the mob when they entered Exeter Change. Their ally, the controversial anatomist Joshua Brookes, was also slandered in the newspapers. Brookes was well known to supply his large anatomy school with corpses through certain unscrupulous "body-snatchers" who robbed fresh graves on the churchyards, and these outrages had, rather understandably, made him somewhat unpopular among the Londoners. There were allegations that he had staged the murder of the elephant martyr only to get the chance to dissect a recently dead elephant. Another rumor stated that Joshua Brookes had cut out several beefsteaks from the elephant's rump, one of which he had eaten for dinner. Even the *Times* repeated this story, but the enraged anatomist, seconded by Mr. Cross, wrote a letter to the editor to deny these allegations and to inquire how a respectable newspaper could carry such calumnies. The editor's only defense was that "the paragraph came from an

Evening paper"! In a popular play, Joshua Brookes was further mocked. It began with a quiet chorus:

If the elephant you'd see
Pray walk in, sir—pray, walk in, sir,
Press among the company,
And dash through thick and thin, sir,
Now you'll see a sight so rare,
Spurzheim and Deville are there,
And Joshy Brooks with arms all bare,
Hacking at the skin, sir!

Later, a chorus of the Exeter Change beasts responded hollowly:

The deed, alas, is done
Accomplished is our fear
Great Chuny's soul is gone.

The cruel Joshua Brookes answers, whetting his scalpel, Yes, but his *trunk* is here! An indignant chorus of men and animals later repels the rapacious anatomist's assault on the elephant martyr's body:

O, Joshua Brooks, turn out—turn out—
Joshua Brooks, turn out—
Do you want to make beef
Of poor Chuny you thief
Fie, Joshua Brooks, turn out!

Another play, entitled *Chuneelah; or, the Death of the Elephant of Exeter Change,* was staged 6 months after the elephant's death, at Sadler's Wells Theatre, with considerable success.

Already in the 1820s, most zoologists considered it cruel and inhumane to keep large wild animals locked up in small cages in an indoor zoo. They harshly criticized Mr. Cross and pointed out the small private zoos of certain British noblemen, and the French Jardin des Plantes, as model establishments of their kind. When, in 1829, it was decided that the condemned old house at Charing Cross was to be razed to provide a site for the National Gallery, Mr. Cross and his animals were homeless once more. He offered the entire stock of the menagerie, with his own services included, to the newly raised Zoological Society of London, but the snobbish zoologists wanted nothing to do with a mere showman. The resourceful Mr. Cross had a surprise in store for them, however. He obtained several impressive backers, among them Queen Adelaide, the Archbishop of Canterbury, and the Duke of Devonshire, who supplied him with money to purchase 13 acres of land at Walworth, a couple of miles outside central London. In 1831, he announced that his new Surrey Zoological Garden would be opened.

When he retired at the age of 70, in 1844, Mr. Cross's new zoological garden was a successful and respected establishment. It boasted a rhinoceros, several beautiful giraffes, a giant tortoise big enough to be ridden by children, many boas and pythons, and an aviary of eagles and vultures donated by Joshua Brookes. Although a catalogue mentions two "dwarf" Asian elephants, there is no record that Mr. Cross ever had a successor to Chunee.

After Mr. Cross's retirement, the Surrey Zoological Garden was taken over by William Tyler, who was one of the men present at Chunee's death. He had worked at Exeter Change for many years and must have been quite an old man. During his reign as director, the zoo went rapidly downhill. Public interest was low, and the animals old and infirm. In 1856, the Surrey Zoological garden was closed and the entire stock of animals sold at auction, often fetching niggardly sums. A group of monkeys could be bought for 10 shillings and two jackals for 24 shillings. It is sad to contemplate that the bears were all purchased by a hairdressing firm, to be made into "bear's grease" for balding pates. The remaining giraffe was purchased by a Continental zoo, but when hoisted on board ship, it fell and broke its neck.

As one is standing at the present-day Strand, surrounded by buses, motor cars, motorcycles, and their noxious vapors, it seems marvelous that this street contained, 160 years ago, a complete menagerie, with lions, tigers, apes, crocodiles, a rhinoceros, and an elephant. Not a trace of the actual Exeter Change building has been preserved to posterity; today, the Strand Palace Hotel is located at its former site. A more apposite monument to Mr. Cross and his beasts can be found in the nearby Trafalgar Square, where the lions at Lord Nelson's column are the work of Sir Edwin Landseer; he is likely to have used his drawings of Nero, the Exeter Change lion, when sculpting them.

Jumbo, King of Elephants

When people call this beast to mind,
They marvel more and more
At such a LITTLE tail behind
So LARGE a trunk before.

Hilaire Belloc, *The Elephant*

THE ELEPHANT LATER TO BECOME KNOWN AS Jumbo, and one of the world's most famous animals, first came into contact with human civilization in 1861, when it was little more than a year old. While taking a stroll along the bank of the Settite River in Abyssinia, the elephant calf was caught unaware by a party of Arab hunters, who soon secured their prey, using their clubs and nets to good effect against the startled animal. Jumbo was, at this time, about 3½ feet tall. The Arabs transported the elephant calf to the coast, where it was sold to the German big game hunter Johann Schmidt. Early in 1862, it was sold to the zoological garden of the Jardin des Plantes in Paris, where it was to reside for 3 years. During this time, nothing unusual was noted about its growth and general behavior. In 1865, an exchange of animals was performed between the Jardin des Plantes and the London Zoological Gardens: the young elephant crossed the channel together with two spiny anteaters, being traded for a rhinoceros. For the elephant, this is likely to have been a fortunate turn of events. During the German siege of Paris in 1871, the starving Parisians raided the zoo for all its edible inhabitants; quite startling dishes were on the restaurant menus at this time, and Jumbo may well have escaped ending up as *ragoût d'éléphant*.

Jumbo at London Zoo

At this time, Mr. Abraham Dee Bartlett was the superintendent of the London Zoological Gardens. He was the son of a hairdresser who had his salon in the Strand, quite near the Exeter Change menagerie. Young Abraham frequently visited Mr. Cross and his beasts at the famous Exeter Change indoor menagerie in Central London, and this early experience was instrumental in arousing his lifelong passion for practical zoology. He became an expert taxidermist, but this trade was not lucrative enough to support him. He was apprenticed to his father and attended to the hair of the living by daytime and the fur of the dead by night. In 1851, he won a prize at the Great Exhibition for a display of stuffed animals and was appointed as naturalist to the Crystal Palace exhibition to take charge of the stuffed animals exhibited among the plants in the enormous greenhouses. The beasts rotted quickly in the humid atmosphere, however, and the appointment was an unhappy one. Bartlett decided to devote his energies to living animals instead of stuffed ones, and he succeeded in obtaining a position as naturalist, and later superintendent, of the London Zoological Gardens. The former hairdresser was very successful as superintendent of the London Zoo. He was quite an eccentric, and his usual dress, even when tending the animals, was a tailcoat and a tall top hat. His skill in handling large and dangerous animals was unsurpassed. Once, he ably sawed off the damaged and malformed horn from a female rhinoceros, in an operation that took 10 minutes, with the animal not objecting in any way. Another time, he pulled a broken tooth from the mouth of a large hippopotamus, using a 2-foot pair of tongs. When two young rhinoceroses were to be moved from one enclosure to another, Bartlett ran before them, his long white beard and coattails fluttering, and dropped bits of bread to guide the rhinoceroses on their way to their new quarters. Not fewer than 24 keepers were pulled after the animals by strong ropes fastened to the beasts' harnesses, in order to slow them down.

Jumbo and Matthew Scott in 1865. From a lantern slide in the private collection of Mr. John Edwards, London. Never before published.

Once Jumbo had been installed at the London Zoo, Mr. Bartlett appointed Matthew Scott, one of the zoo's veteran keepers, to take care of the young elephant recently received from France. Scott had become something of a hero a few years earlier, when he had lured an escaped hippopotamus back into its enclosure. The ungainly beast charged him at a surprising pace, and the short-legged keeper had to run for his life. He scrambled over the outer fence of the paddock just in time to evade the gaping jaws of the enraged animal. Matthew Scott was normally a keeper of antelopes, but the eccentric Bartlett considered this an advantage, since he would be more attentive to the directives of the superintendent if he lacked preconceived ideas of his own about how to handle pachyderms. The young elephant needed a lot of attention during its early stay at the zoo, since it had not been well cared for in Paris: its hide was ingrained with filth, and its feet were overgrown since the nails and hard calluses had been allowed to develop without any attention. Mr. Bartlett and Matthew Scott took good care of their young charge. The hide was washed and brushed, the feet filed and pared, and nourishing meals administered three times a day.

The elephant's name, Jumbo, is likely to have been chosen by Abraham Bartlett. It had not been previously used for elephants. The word *jumbo* had belonged to the slang language since the early 1800s, being used to mock clumsy, thick-set, and corpulent individuals. The elephant was thus given its name owing to its (presumed) clumsiness rather than its size. According to another interpretation, the elephant's name was derived from "mumbo-jumbo," the name of a powerful African supernatural being. The adherents of this theory are supported by the fact that Bartlett later named an African gorilla Mumbo. There are also several other derivations of the word that might explain its use as the name of an African elephant: the Zulu word *jumba* means a large packet, for example, and the Angolan vernacular name for elephant is *jamba*. It is remarkable to note that, according to the description of the Tower menagerie published in 1800, a large baboon was named Jumbo. It is unknown whether it was just a coincidence that another outsized African animal in a London menagerie was given the same name over half a century before Jumbo was born.

Jumbo was the London Zoological Gardens' first African elephant (*Loxodonta africana*). The elephants of this species are considerably larger than their Indian cousins (*Elephas maximus*) and are further distinguished by their huge ears and by the cloven end of their trunks. Mr. Bartlett was well aware that African elephants were considered to be more temperamental than Indian ones and that some zoologists considered them wholly untamable. From the very beginning, however, he was favorably impressed by Jumbo's intelligence and great affection for his keeper. Since Jumbo needed his daily exercise, the elephant was used to give rides to little children, who traveled in a large howdah strapped onto the elephant's broad back. Jumbo seemed like an ordinary African elephant until the age of 7. At that time, Matthew Scott noticed that Jumbo's appetite increased dramatically: the elephant daily consumed 200 pounds of hay, 2 bushels of oats, a barrel of potatoes, several quarts of onions, and between ten and fifteen loaves of bread. Matthew Scott often gave Jumbo a bucket of strong ale and sometimes even a bottle of whisky. Some teetotalers were concerned about this, but Matthew Scott claimed that the whisky was an elixir of health for Jumbo, and that the ardent spirits aided his prodigious growth.

Jumbo grew in proportion to his intake of food, and he was soon Europe's largest elephant. Only an Indian elephant belonging to a maharajah was alleged to surpass him in size. For many years, Jumbo was the foremost attraction of the London Zoo: innumerable children, including the Prince of Wales, had ridden on his broad back. Young Theodore Roosevelt saw Jumbo when he visited London, and Winston Churchill had his picture taken with the famous elephant.The elephant rides were supposed to be free, but it was good manners to hand a shilling or two to Matthew Scott when Jumbo returned. If this sum is multiplied by 12 (the number of children Jumbo could carry in his howdah) and then by 4 (the number of elephant rides each day), it is easy to understand why Matthew Scott wanted

Some officials of the Zoological Society of London taking a ride on Jumbo in the late 1870s. The man in a gray bowler hat holding the elephant's trunk is Matthew Scott. From an old photograph in the author's collection.

to keep his sole rights to the elephant riding at the zoo. Matthew Scott's prestige at the zoo increased with the fame of his charge; he was soon known as "Jumbo's keeper," and at the pub, the other keepers respectfully made way for him when he went in to have a pint of bitter to refresh himself in between elephant rides.

Throughout the 1870s, Jumbo grew at a steady rate. He was established among the "sights of London," and the favorite of the royal family. Queen Victoria never failed to visit him while at the zoo and give him a loaf of bread. In 1880, Jumbo was nearly 11 feet tall at the shoulder and weighed 6 tons; with his 7-foot trunk, he could reach objects 25 feet above ground level. That year, the previously docile elephant showed his first signs of temper: he gored the iron cage doors with such force that both tusks were broken near the jawbone. The tusks then grew into the jaw, and a painful abscess formed. The clever Mr. Bartlett made the correct diagnosis, however, and ordered a long harpoon to be sharpened. Together with Matthew Scott, he went into the elephant house and, without warning, thrust the harpoon into the swollen jaw, causing the effusion of a large amount of very offensive pus. Jumbo uttered a tremendous roar but did not attack them. The next day, Bartlett could incise the other abscess in the same way, without the elephant objecting in any way.

After these successful operations, Jumbo's health improved rapidly, and his broken tusks once more grew through their proper apertures. For several years to come, however, he periodically became aggressive and difficult to handle. No one but Matthew Scott had the power to calm Jumbo down during these paroxysms of rage; as soon as he went into the elephant den or took Jumbo for a walk in the gardens, the elephant's fury abated. Matthew Scott was a stubborn, difficult character, however. He persistently refused to let any other person handle Jumbo, and particularly objected to having a deputy elephant keeper under him, since this would have meant that he had to share the earnings from the lucrative elephant rides. In his youth, Abraham Bartlett had been one of the horrified witnesses to the violent and bloody destruction of the infuriated elephant named Chunee in the Exeter Change indoor menagerie. With these tumultuous happenings in mind, he gave orders to purchase an elephant gun, to be used if Jumbo ran amok at some time when Scott was away. Mr. Bartlett believed that Jumbo's occasional paroxysms of fury were due to *musth,* the periodic insanity of bull elephants during the rutting season. This diagnosis is by no means impossible, since Mr. Bartlett was an expert zoologist with previous experience of elephants. Some zoologists have advanced another theory, however, after studying casts of Jumbo's teeth. In the upper jaw, the teeth are abnormally developed and severely maloccluded, and Jumbo's periods of fury started just as the fifth molars were erupting in each jaw.

P. T. Barnum Interferes

After Phineas Taylor Barnum's American Museum had been destroyed by fire in 1865, this clever and active showman entered the circus business with considerable success. After joining forces with his colleague James A. Bailey, he was the owner of America's leading circus. Barnum & Bailey's Greatest Show on Earth employed 370 circus performers and had a huge menagerie of 20 elephants, 338 horses, and 14 camels, in addition to many zebras, lions, leopards, hyenas, and large snakes. For some time, Barnum had coveted Jumbo, but he had considered it highly unlikely that the Zoological Society of London would even consider selling this famous elephant. In January 1882, Barnum's London agent once more approached Mr. Bartlett about Jumbo. Initially, the London zoologists were reluctant to sell Jumbo, but when Barnum offered 10,000 dollars (2,000 pounds at the time) for the elephant, they changed their minds and promised that the board of the Zoological Society would debate the matter. At first, not even Barnum believed that the deal would go through, but after 2 days of debate, the zoologists decided, by a small majority, that the elephant was to be sold. The cause was officially stated to be Jumbo's uncertain temper, and it was also claimed that his size was making him impossible to ride. Many people later presumed, however, that the considerable sum of money offered was instrumental in prompting the sale, since the financial state of the venerable Zoological Society was not strong. Because of Jumbo's immense popularity, the zoologists were unwilling to break the news to the press and the British public, who would not, they knew, take it kindly that their great favorite was being shipped to America. The news leaked to the newspapers, however, probably through the agency of the publicity-minded P. T. Barnum, and a national furor arose when it became known that Jumbo was to be sold.

Many Britons at first believed this tale of Jumbo being sold to a vulgar American to be some kind of tasteless joke, but the secretary of the Zoological Society of London reluctantly had to confirm to the press that Barnum had really bought the elephant. This "outrageous sale of a national character" disgusted many people whose children had once ridden on Jumbo: hundreds of vituperative letters flew to the offices of various newspapers. The children themselves wrote sentimental letters to the queen or to Prime Minister Gladstone, begging that the wicked Barnum not be allowed to take their beloved Jumbo away from them. Queen Victoria, the Prince of Wales, and the aesthete John Ruskin together wrote an appeal to the Zoological Society, declaring themselves to be astonished at the actions of this society and urging the zoologists not to let Barnum have Jumbo. The queen would pay any costs incurred by this breach of contract.

Despite his Yankee bravado and businesslike utilitarian attitude, P. T. Barnum had always had a secret admiration for Britain—its glorious past, historical monuments, and famous writers and poets. Although some of his biographers have depicted him as a coarse, illiterate, boorish man, Barnum, albeit no scholar, was well read in the English classics and even staged Shakespearean dramas at his American Museum. His admiration for Britain and its national treasures sometimes inspired him to attempt business deals that clearly did not amuse the jingoist Britons, who viewed this exuberant Yankee with great skepticism. Barnum once very nearly succeeded in purchasing the cottage in which William Shakespeare was born, at Stratford-upon-Avon. He intended to have it carefully dismantled and shipped to New York for reassembly. At the very last minute, a consortium of public-minded Englishmen managed to outbid him, turning the building over to the Shakespeare Trust. According to rumor, another audacious transaction nearly deprived London of one its prime attractions, Madame Tussaud's Museum at Baker Street, which narrowly escaped being shipped to America. Allegedly, Barnum once even bid 2,500 dollars for a famous oak tree, on which Lord Byron had once carved his name, but his offer again fell through. The furor caused by these attempts to remove traditionally British monuments was nothing, however, to the nationwide fury caused by P. T. Barnum's attempt to kidnap Jumbo to the United States. Lord Winchilsea must have expressed the opinion of many Englishmen when he suggested that the prime minister, William Gladstone, was to be sold to America in Jumbo's stead:

But since in England's fallen state
She owns two things supremely great,
Jumbo and Gladstone — (each we find
The most prodigious of their kind) —
And one won't budge. The, Barnum, make
A fair exchange, for quiet's sake!

Take the Right Honourable, and go!
He'll make the better raree show!
Leave Jumbo.

Jumbo-Mania

Next to P. T. Barnum himself, the obvious target of abuse for patriotic Englishmen was Mr. Sclater, secretary of the Zoological Society of London. After surreptitious attacks in the press, this gentleman had to write a retort to the virulent criticism directed against the society. He emphasized that Jumbo had, in later years, seemed excitable and easily enraged; several experienced zoologists had feared that Jumbo would have had to be shot if he had gone berserk within the zoo. Mr. Sclater pointed out that this, and not the monetary aspect, was the cause for Jumbo's being sold. In America, Jumbo would not be used as a riding elephant but would only be shown during circus performances. Mr. Sclater was particularly offended by the allegations from several journalists that Jumbo would be ill treated in the United States. He pointed out that Barnum & Bailey had more than twenty elephants in their circus and that one of them had recently had a calf, something that happened only rarely among elephants in captivity. Mr. Sclater was sorry to announce that both he and Mr. Bartlett had received many threatening letters, in which they were, with suitably zoological metaphors, called "skunks," "reptiles," and "craven beasts." Poor Matthew Scott had, by another earnest letter writer, been likened to Judas Iscariot, since he had basely betrayed Jumbo to his enemies.

Mr. Sclater's defense did not impress Jumbo's many friends. If Jumbo was dangerous and vicious, why then had he been allowed to consort with young children just a few days earlier? And if Jumbo was dangerous, would it be morally correct to sell him off to the United States, thereby valuing American children's lives lower than English ones? The editor of the *Daily Telegraph* sent a telegram to Barnum in the name of the British nation:

> Editor's compliments. All British children distressed at elephant's departure. Hundreds of correspondents beg us to inquire on what terms you will kindly return Jumbo.

Barnum immediately sent a telegram in return, but its contents did not please Jumbo's many friends:

> My compliments to Editor Daily Telegraph and British nation. Fifty millions of American citizens anxiously await Jumbo's

> arrival. My forty years' invariable practice of exhibiting the best that money could produce makes Jumbo's presence here imperative.

It thus seemed as if Jumbo could not be saved for the British nation. The *London Standard* compared the act of separating Jumbo from his British friends to a southern slave owner selling the members of a black family separately at auction. In the *London Telegraph,* the avaricious American was solemnly cursed, and poor Jumbo's future among the Yankees was bemoaned:

> No more quiet garden strolls, no shady trees, green lawns, and flowery thickets. Our amiable monster must dwell in a tent, take part in the routine of a circus, and, instead of his by-gone friendly trots with British girls and boys, and perpetual luncheon on buns and oranges, must amuse a Yankee mob, and put up with peanuts and waffles. . . . We fear, however, that Jumbo will never come back to us alive. His mighty heart will probably break with rage, shame, and grief; and we may hear of him, like a another Sampson [sic], playing the mischief with the Philistines who have led him into captivity, and dying amid some scene of terrible wrath and ruin.

In the *Vanity Fair* weekly, the editor started a fund for the preservation of Jumbo. He himself donated 5 pounds and wrote that since an elephant could live to more than 120 years, Jumbo was likely to survive the decline and fall of the British Empire. Although the editor's political intuition was apparently more highly developed than his zoological knowledge, the readers of *Vanity Fair* were quite interested and contributed considerable sums of money, although far from enough to outbid P. T. Barnum. A poor old lady, who had no money to give, instead donated her sewing machine to save the elephant.

Jumbo appreciated his many visitors, and particularly the buns, cakes, and fruit baskets they had brought with them. People had rather confused ideas of what an elephant liked to eat, however. After an inspector of the Royal Society for the Prevention of Cruelty to Animals had stopped an individual who tried to feed Jumbo a leg of mutton, one of the keepers was posted on guard outside Jumbo's den to stop people from giving him kippers, bones, beefsteaks, oysters, whipped cream, or other food wholly unsuitable for an elephant. An old lady bringing Jumbo a box of sweets and hothouse grapes was confounded when the elephant instead seized a basket of flowers, which he was meant to smell, and devoured it whole, ribbons and all. Once, a kind-hearted visitor gave a loaf of bread to the completely neglected Indian elephant in the den next to Jumbo's, but the huge neighbor put his head into the den, stretched out his long trunk, and seized the loaf. Every day, a well-dressed old lady stood in front of Jumbo's den, distributing handwritten pamphlets containing a prayer that the Lord would forgive the heretical scoundrels who had sold Jumbo to the American barbarians, and that He would intercede to stop Jumbo being moved to the United States. Three times a day, she knelt opposite the elephant house to personally present the elephant's case before the Lord.

Thousands of buns were sent to Jumbo by children all over Britain. Other individuals sent the elephant an amazing variety of gifts: fruit, sweets, puddings, cakes, and sandwiches arrived at a steady rate, and a gardener sent a huge prize pumpkin, which Jumbo devoured with relish. Eccentric individuals sent bottles of wine, champagne, beer, and whisky and cases of snuff and cigars; Matthew Scott cautiously stated to the press that these gifts "were put where they would do more good"! Another well-wisher sent a huge nightcap to Jumbo, intended to prevent his catching a cold while crossing the Atlantic. Another eccentric gift was a huge crate of pills against seasickness. Alice, the African cow elephant known as "Jumbo's wife," although the two animals seldom met and never showed interest for each other, was given a widow's cap to wear while mourning her departed husband. Innumerable letters were addressed to "Jumbo, Zoological Gardens." Most of them were from children, containing sentimental verses or phrases like "Jumbo don't go" or "Jumbo come home Christmas." Many children sent Jumbo locks of their hair; they probably imagined that Jumbo would treasure these mementos of his little friends and paste them into an album along with the sentimental poems. Another letter came from an (apparently quite sane) clergyman, who advised the elephant "to make a firm fight of it!"

In the London shops, Jumbo hats, Jumbo coats, Jumbo cravats, and Jumbo fans had a roaring trade. At the restaurants, Jumbo soup and Jumbo stew were served, although neither of these dishes was actually made from elephant meat. The color à la mode for the satin dresses of the fashionable ladies was a dark grey, also named after the famous zoo elephant. Many books, broadsheets, and music books featuring Jumbo were printed by clever entrepreneurs and avidly bought by the public. In the *London Fun,* it was suggested that the lion in the British coat of arms be changed into an elephant, with the accompanying device *Dieu et Mon Jumbo!* Americans were quite unpopular in London during this time, and the U.S. ambassador James Lovell remarked in a speech that "the only burning question between the two nations is Jumbo."

But although the odds were against them, Jumbo's British friends did not give up. Egged on by the royal family and by several influential politicians, a group of dissidents within the Zoological Society of London, led by Sir George Bowyer and Mr. Berkeley Hill, asserted that the sale of Jumbo was in disagreement with the original charter of that society. Not only was it morally wrong to sell a dangerous animal, but the sale of this huge elephant, a type specimen of its kind, to a mere showman would seriously hinder the study of zoology. They took the Zoological Society to court, and the trial began before Mr. Justice Chitty on March 9, 1882. All the newspapers published lengthy accounts of this trial, and the Londoners avidly read the pro-Jumbo editorials. The Zoological Society soon gained the upper hand by conclusively proving that, according to their charter, the board had the authority to sell any animal without consulting all the fellows. Sir George Bowyer compared the sale of Jumbo to a similar outrage affecting the British Library: what to do, if the governors of this venerable institution would sell the Magna Charta or the Codex Alexandrinus to the highest bidder? He argued that Jumbo was just as important for the study of zoology as these documents were to the study of history, and that the Society had no right to sell "any article valuable for the study of natural history." The zoologists again objected that they had sold animals before, in particular a gnu for 750 dollars. The dissidents replied that although less valuable animals could be sold, Jumbo was a "type specimen" of what was presumed to be a newly discovered, larger form of African elephant. Another argument was made that, at this time, African elephants were heavily decimated by the ivory trade, and many zoologists feared that they soon would only be met with in zoological gardens. Mr. Justice Chitty's verdict was that the sale of Jumbo broke no law and that the elephant now was the property of P. T. Barnum. In his summing up, the judge lamented that during this trial, he had received an unprecedented amount of sentimental and threatening letters from the public. Like the Editor of the *Times*, Mr. Justice Chitty hoped that this final sentence would end the Jumbo mania that had affected all Londoners.

Jumbo Leaves the Zoological Gardens

Together with Matthew Scott, Barnum's agent tried to lure Jumbo into a cage with wheels under it, inside which the elephant was to be pulled down to the London docks. They had considerable difficulties in handling the elephant, however, since Jumbo proved to be quite restive and easily frightened once he was removed from his habitual surroundings inside the Zoological Gardens. He was very frightened by horses and sat down and remained seated as long as any member of the equine race remained in view. Barnum's agent reported to his boss that the morose little Matthew Scott was far from cooperative. Barnum replied that help was on its way: as soon as the Zoological Society had agreed to sell the elephant, William Newman, one of Barnum & Bailey's most skillful elephant keepers, had been dispatched across the Atlantic to bring Jumbo back with him. After Newman and Scott had made plans, another attempt was made to lure Jumbo into the cage, which resembled a very large horse box. Although Jumbo was put in chains by Newman, he refused to mount the ramp up to the crate. The next day, when Jumbo was led toward the cage, he lay down on the ground and refused to go near it. The spectators applauded the elephant and sang "He's a jolly good fellow!" "Elephant Bill" then brought out his long pike staff, which he used to discipline his elephants back in the United States, but the local representative of the Royal Society for the Prevention of Cruelty to Animals interceded on Jumbo's behalf and forbade him to use this cruel weapon. Jumbo chose to lay down just outside the parrot house of the London Zoo, whose denizens let out a communal shriek when they were excited by Jumbo's trumpeting, the curses of the keepers, and the hearty cheers of the spectators. The other zoo animals also contributed various sound effects, and even the stolid elephant cow Alice trumpeted loudly.

As soon as the furious Newman had given up and ordered that Jumbo was to taken back to his enclosure, the elephant promptly rose at Matthew Scott's command and plodded back to the zoo in his footsteps. The journalists threw themselves into their hansom cabs and rushed back to the headquarters of their papers to report the latest news: as a true Englishman, Jumbo refused to go to the United States, and his faithful wife Alice wept and called for him to come home! Barnum's agent wired his boss to inform him that Jumbo would not budge, but the clever showman replied that he was allowed to lie there as long as he wanted to—"it is the best advertisement in the world." "Elephant Bill" Newman was less amused by Jumbo's peccadilloes, however, and according to a (probably exaggerated) newspaper report, he had even declared that he had the right to shoot Jumbo if this proved necessary: "Living or dead, Jumbo is to go to New York!" This declaration is unlikely to have added to his popularity among the Londoners.

These dramatic occurrences at the zoo were given as much attention in the newspapers as a recent attempt to assassinate Queen Victoria. This sentimental story of the poor elephant objecting to be exiled to the United States—and triumphing, as an honest British underdog, over the Yankee vulgarians sent to remove him—arose much interest. One Londoner sent a wreath of flowers to Jumbo "as a trophy of triumph over his brutal owners and American kidnappers," and a fashionable young lady who had just married sent half her wedding cake as a present to the elephant. One newspaper leader compared the sight of Jumbo in chains, in preparation to be delivered to the Americans, to that of Mr. Selby disposing of the honest, uncomplaining Uncle Tom. An eccentric individual christened his son Jumbo, but the religious press urged that he was to be fined for this outrage, since with such a name, the poor boy would be condemned to be a "rough" for life. Entire schools of children petitioned the Zoological Society, and Barnum claimed that hundreds of children had written to him, begging that Jumbo be spared. He reproduced one of these pathetic letters, allegedly written by "A young English girl," but the style and Americanized spelling would suggest that it had been written by himself or by some unscrupulous journalist in his employ!

A sentimental poem, written at the height of the Jumbo craze and illustrated by a drawing of Jumbo and Alice at the Zoo, contained the following words:

When quite a baby I came here, and now to London folk I'm dear
They'll try to keep me yet, I know, from Barnum and his travelling show.
It grieves me sadly to be sold for just two thousand pounds in gold,
And could I talk I'd quickly say, 'I'm treated in a shameful way.'
They chained me up one day, to be shipped across the raging sea,
But I, your faithful friend Jumbo, did not just feel inclined to go.
Again they tried the nasty chain, but all their efforts were in vain,
For with a very angry frown upon the ground I laid me down.
I love the brave old British flag, of it, my boys, I'll always brag,
And you must clearly understand, I do not care for yankee land.
Leave me with Alice kind and true, leave us together in the Zoo,
And let our friend Squire Barnum know, I can't go with him in his show.

A young boy, writing to *Vanity Fair,* was very much disappointed with the elephant cow Alice, who "did not look miserable at all" at the prospects of losing her dear husband.

The editor of *Punch* magazine did not take part in the rampant Jumbo-mania. He instead pointed out that while ample funds were available to keep Jumbo on the right side of the Atlantic, a mission dedicated to giving a dinner of Irish stew to the starving children of the East End on every Wednesday was failing owing to lack of funds. In the accompanying poem, a forthright East Ender gives his views on the Jumbo controversy, contrasting the lot of the pampered elephant with that of the stunted, filthy children of his own class:

> *We can feel the gnawing hunger and* we *never gets our fill,*
> *No columns in the TELEGRAPH when TOM or DICK falls ill.*
> *There's no national subscription to keep us over here —*
> *No! It strikes me they're uncommon glad when 'outward bound we steer.*
> *Bu then* we're *not all elephants, we're only rags and bone,*
> *To be gathered by the dustman, and left unfed, alone;*
> *To be cast upon the gutter, and to grovel in the slums;*

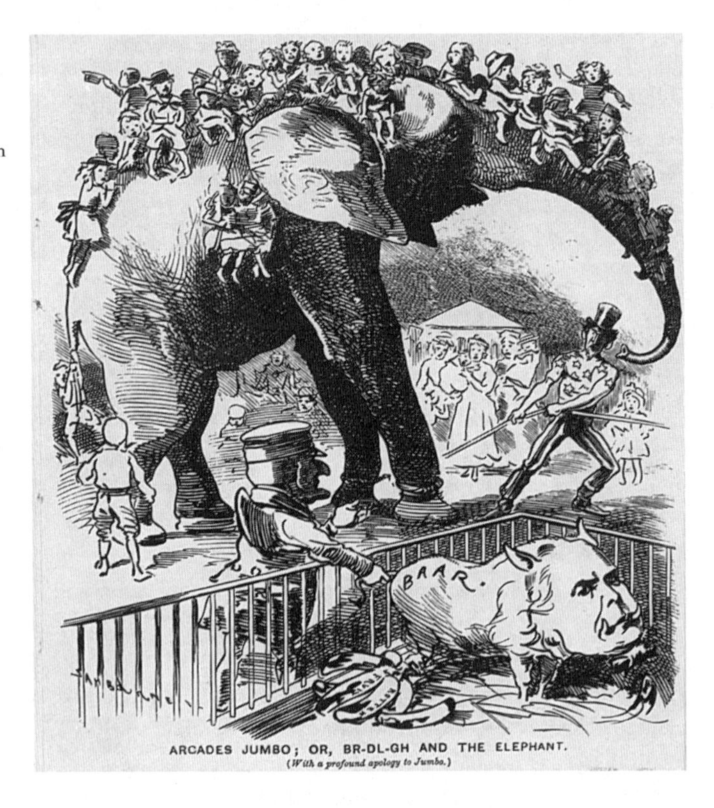

The caricature *Arcades Jumbo*, published in 1882. Mr. Punch suggests to P. T. Barnum that he should bring the politician Charles Bradlaugh, who is depicted in the guise of a pig, with him to the United States, instead of Jumbo, whose back is cluttered with grieving British children. From a print in the author's collection.

To never have a decent meal, and raven for the crumbs;
To take to lying and to theft, to blasphemy and curse,
Maybe to fill the prison cell, maybe to leave it worse.

The clever Mr. Bartlett had carefully observed the futile attempts to get Jumbo into the cage. He had seen that at the same time Matthew Scott called out to the elephant to rise, he also clandestinely made signs to his great charge to remain in a recumbent position. The sly old "Scotty" had wanted to keep his old job as elephant keeper at the zoo, to earn ample money from the elephant riding, and hobnob with the nobility and gentry wanting to see Jumbo. His immediate rapport with Jumbo had been noted by Mr. Bartlett many years earlier, and it had been further strengthened as the years went by. Scotty could direct Jumbo only by signs and gestures. He had played the fool before the Americans, who had not seen through his deception. Mr. Bartlett called Matthew Scott to his office and reproached him for his disloyal behavior, which was seriously threatening the zoo's agreement with P. T. Barnum. Bartlett saw no possibility of keeping the disobedient Mr. Scott within his establishment if he did not agree to accompany Jumbo to the headquarters of Barnum & Bailey's circus in the United States. There, he would keep his position as Jumbo's keeper, at least for the time being, and be paid a good salary. Although Matthew Scott was dismayed to be put under the command of an American, he grudgingly agreed to this offer. In the meantime, Elephant Bill ordered a novel kind of wheeled elephant cage, which had two large gates that could be slammed shut if needed, to be built by a company in the American Midwest and shipped across the Atlantic. It arrived just in the nick of time, when Jumbo had become considerably more docile after Scott had ceased his subversive actions. The elephant could be led through the crate on several occasions, and when the time was right, the two gates were slammed shut and Jumbo's legs were chained. For an hour, the elephant struggled furiously to get free; the whole cage shook, and each time a plank was torn loose, a rowdy mob outside cheered and exulted. It was feared that Jumbo would be able to tear the chains asunder, but Barnum's strong ironware was hardy enough to confine even a furious elephant. Elephant Bill and his colleagues triumphed, and they could finally nail a sign with the address "Barnum & Bailey, New York, USA" to the crate—this probably more owing to advertising purposes than the fear that it would be sent to the wrong address.

Ten great draft horses were hitched onto Jumbo's crate and pulled the elephant out into Regent's Park without difficulty. The London zoologists had warned Elephant Bill that the wheels of the trolley were likely to be too narrow to carry Jumbo's bulk over the graveled walks of Regent's Park, but the American had not heeded their advice. As soon as the elephant was hauled into Regent's Park, the wheels sank into the earth. Only through jacking and digging the wheels out at regular intervals could Jumbo be kept moving. When Jumbo's crate was negotiating a steep, grassy mound near Regent's Canal, the police and zookeepers guarded its path and egged the horses on, dreading a stop at this point where the ground was exceedingly soft. Just as they thought they were clear, some mischievous person in the crowd loudly cried out "Whoa!", and all ten horses obediently stopped short. Before the team of horses could be persuaded to get moving again, the axles and wheel had sunk into the soft earth of the Regent's Park walk. It took many hours to dig them out and to lever planks under the wheels, and it was not until just after midnight that Jumbo's crate finally moved forward again. In spite of the early hour and the cold, drizzly weather, a rough crowd had assembled before the gates of the Zoological Gardens, singing "Rule Britannia" as Jumbo was slowly approaching, pulled by the ten big, steaming horses.

The progress was swift through the silent streets of sleeping London. At regular intervals, the party had to stop to refresh the horses and to pour water over the steaming wheels. Elephant Bill and Matthew Scott, who rode on a platform in front of the elephant's crate, were probably grateful that Jumbo was moved just after midnight. Had such a conspicuous procession taken place by daytime, the elephant's progress would inevitably have been impeded by a large and rowdy mob, and Jumbo's attendants would have been in immediate danger of lynching. Even at this late hour, a crowd of two hundred people were running along with the elephant crate, and a few "bloods" and newspaper men were pursuing it in their hansoms. At the military barracks of Albany Street, the guard turned out and shouldered arms to the passing elephant. Initially, Jumbo took things quite calmly, comforted by Scott, who stood just before the elephant, fondling the trunk. At one occasion, however, Jumbo stretched out his trunk and tweaked the tails of the horses. Later, he took hold of the reins of the cart, to the delight of the crowd, who shouted out that Jumbo was driving himself!

At 7:00 A.M. the whole company had breakfast, and Jumbo was given two dozen oysters and a bowl of champagne by an eccentric peer. Once at St. Katherine Docks, Jumbo's crate was unloaded and hoisted over to a large Thames barge. Here, a lady friend of Jumbo, who had traveled many miles to take a fond farewell of the elephant, was allowed to pour two quart bottles of beer into the trunk, and feed the elephant many buns and cakes. Later in the morning, Jumbo's crate was hoisted onto the *Assyrian Monarch,* a large steamer used to ship immigrants to the United States.

Once on board, Jumbo had many visitors. Barnum's British agent staged a gala luncheon on board, to which many noblemen, politicians, socialites, and newspaper men were invited. All of them went to see Jumbo afterward, and the American consul general gave a speech, expressing his hope that the anti-American sentiments that had been aroused by the untimely selling of the elephant would abate as soon as Jumbo was freighted across the Atlantic, and that the great elephant would one day unite Britain and the United States in unanimous sympathy. Elephant Bill was, at the same ceremony, awarded the gold medal of the Zoological Society of London for his skillful handling of the elephant during its transport from Regent's Park to the Docks. As Jumbo's last visitor on British soil, the Baroness Burdett-Coutts boarded the *Assyrian Monarch,* along with a distinguished party of noblemen and ladies. She gave the elephant a last bun and shook hands with his keepers Scott and Newman. No one noticed the four hundred poor Russian Jews who had bought their cheap third-class passage on the *Assyrian Monarch;* compared to Jumbo, they might as well have been fleas on the coat of a prize dog. Nor did any one care about the two hundred further emigrants who had been displaced by the elephant. The poor wretches craned their heads to see the huge beast embarking, apparently feeling much honored to travel in the company of such an international celebrity. When the Baroness Burdett-Coutts graciously spoke to some of the children, they knelt down before her and kissed the hem of her gown, believing her to be Queen Victoria who had come to say farewell to her beloved Jumbo.

When the *Assyrian Monarch* hove to at Gravesend, the men at war manned the rails to honor the elephant. At every pilot cutter in British waters, communiqués about Jumbo's state of health were carefully deposited. As the ship lay off the Nore, an entire bottle of whisky was poured into Jumbo's trunk as a celebration. The elephant was mightily seasick during the first few days afloat, since the ship was struck by a storm while out into the Atlantic. At many London churches, prayers were said in favor of animals and men in peril on the sea, after some disreputable rags had spread the rumor that

Jumbo's ship was in danger of sinking. The storm calmed down, however, and after a 15-day journey, the *Assyrian Monarch* lay to in New York; during the following months, the Jumbo-mania slowly abated in London and the British provinces.

Jumbo as a Circus Elephant

When Jumbo went ashore in New York on April 10, 1882, P. T. Barnum was at the quay to receive the most recent—and valuable—addition to his circus. He was deeply touched when he saw the gigantic animal, which he had paid 30,000 dollars to purchase and transport across the Atlantic. In meandering detail, he described to a crowd of newspaper men how he had, several years earlier, ridden on Jumbo's back together with his tiny protégé, the famous midget General Tom Thumb. Some of the journalists, themselves already the worse for drink, decided to enliven the occasion by pouring a couple of bottles of whisky down Jumbo's capacious trunk. P. T. Barnum objected to their intoxicating his elephant, but the merry newspaper men did not heed his warnings and continued their horseplay. One of them admiringly remarked that Jumbo was able to swallow down two large bottles of whisky in one single gulp and to drink a bucket of beer as a "chaser," just like a human being. During the welcome ceremony that had been arranged for Jumbo, the elephant was pulled through Broadway in a cart pulled by sixteen horses and pushed by two other elephants. P. T. Barnum's three brass bands played rousing music at a deafening note as the parade was approaching the Madison Square Gardens, and the inquisitive and obtrusive mob was frightening poor Jumbo, who was used to his tranquil life at the London Zoological Gardens, out of his wits.

In the circus flyers and handbills, Jumbo's proportions were greatly exaggerated: "His trunk is the size of an adult crocodile, his tail is as big as a cow's leg, and he makes footsteps in the sand of time resembling an indention as if a very fat man had fallen off a very high building." It was also hinted that Jumbo had nearly caused a war between Britain and the United States, and that the dastardly Brits had attempted to feed Jumbo poisoned buns just before his removal to America, believing that their darling elephant would prefer death to slavery among the vulgar Yankees. A witty columnist for the *New York Times* decided to improve on Barnum's flights of fancy. He remarked on P. T. Barnum's remarkable self-restraint in not mentioning several remarkable facts about the elephant's life in London. It was well known in court

Scenes from the Life of Jumbo, a broadsheet from the private collection of Mr. A. H. Saxon, reproduced with permission.

A coach drives under Jumbo's belly in this exaggerated advertisement.

circles, he asserted, that from her earliest years, Queen Victoria played and romped with Jumbo at the Windsor Castle Park, making the elephant fetch and carry like a dog, and rolling and tumbling with him on the turf. Later in her life—when the royal veterinary advisers tactfully pointed out that the risk of her Majesty accidentally rolling on Jumbo and seriously injuring the poor animal was too obvious to be disregarded—the stoutly built queen had to be satisfied with keeping Jumbo indoors like a well-trained poodle, and to have him sit at her side and "beg" for sugar lumps at teatime. After Lord Beaconsfield had procured for the queen the title of Empress of India, she became very fond of riding Jumbo around the backyard of Buckingham Palace. Lord Beaconsfield used to sit on the elephant's neck to act as a mahout. This practice was, of course, carefully hidden from the liberal statesmen of the time. It might never have been known outside the palace, had not Jumbo entangled his trunk with one of the royal clotheslines one Monday morning and bolted across the yard, throwing both the empress and the mahout. This spirited narrative has no semblance of truth whatsoever, and can be considered an important forerunner of the reporting of the present-day antics of the British royal family in the popular press. P. T. Barnum was delighted with this clever spoof of his own advertising pamphlets. In a letter to the editor, he replied that it could now not be doubted that all the stories about Jumbo's loving intimacy with Queen Victoria were true—"We all know that the Queen's attachment to Jumbo was so ardent that for years there could be no sleep on the Royal couch until Jumbo had sounded his trumpet 'good night' from the Windsor Park or the back yard of Buckingham Palace. . . ."

After some months, Jumbo became used to circus life. When Barnum & Bailey's Greatest Show on Earth performed, Jumbo took precedence over all other two- or four-legged artists. There was a general hush when the ringmaster shouted out,

> The Towering Monarch of his Mighty Race, the Colossus of Elephants, The Biggest and Most Famous Animal in the World, ridden upon by Queen Victoria, the Royal Family, and more than a million Children—just arrived from the Royal Zoological Gardens, London—JUHUMBO!

As the line of nearly thirty elephants trudged into the ring, led by the mighty Jumbo, the crowd was dazzled by their size and power. At the tail of the parade went two elephant calves, which were the objects of much sentimental affection. Jumbo usually walked together with a small "clown elephant" called Tom Thumb, to further emphasize his great size. Tom Thumb later performed a comic show with the circus clowns, pretending to steal bottles of liquor from them and get more and more drunk. Other elephants stood on their forelegs, played on the seesaw, or balanced on barrels. Jumbo himself was too old a "dog" to learn new circus tricks, and during the actual shows, he dignifiedly stood to one side, watching the cavorting of the other elephants with an air of slight boredom.

Jumbo was the largest, as well as the most famous, immigrant into the United States. As in Britain, the elephant became quite a media personality. Cigars, crocheted potholders, and a particularly big sewing machine were all named after Jumbo. The use of Jumbo in advertising was quite extraordinary, even by American standards. Baking powder, sewing thread, soap and tooth powder were advertised using the famous elephant's image on the trade cards. An astonishing drawing from this time depicts Jumbo stampeding through a desert landscape, wearing patent boots with clinching screw fasteners. The caption reads "The two biggest things on earth: the Clinching Screw and Jumbo!!!" The sun shines benignly on the great, well-booted elephant as he runs by, carrying a banner emblazoned with the words "For Hard Work & Constant Use give me the 'Clincher.'" A party of turbaned Arabs reverently stands aside when the elephant pounds by them in his ungainly footwear. Another, even more far-fetched advertisement depicts Jumbo feeding an elephant calf castor oil, as an advertisement of the laxative Castoria; it is accompanied by the following deplorable poem:

From peasant nurse to high-born lady
All mothers know what's good for baby, CASTORIA.
While Jumbo, too, though not a lady
Follows suit and feeds the great baby CASTORIA.

Jumbo proved to be one of P. T. Barnum's most shrewd investments ever. In just 2 weeks, the elephant had repaid the 30,000 dollars consumed by its purchase, transport, and advertisement campaign, with a 20 percent interest. In his first season at the circus, Jumbo was taken on a 31-week tour from coast to coast, which earned 1.7 million dollars. Jumbo traveled in a specially built railway car called "Jumbo's Palace Car." It was richly decorated in red and gold and was mounted on two six-wheeled bogies, with double entrance doors in the middle, and reached almost down to the track. Matthew Scott had his quarters in the same railway car, and when on tour, both he and Jumbo lived, ate, and slept there. There is a story that, just before getting into his bunk at night, Matthew Scott liked to drink half a large bottle of beer

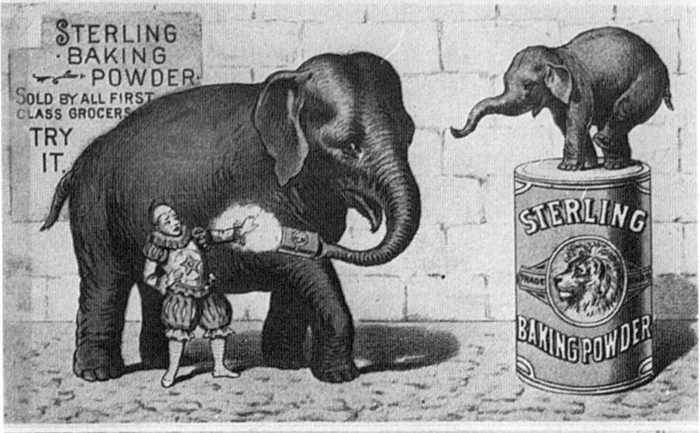

Three contemporary advertisements, using the famous Jumbo to puff some popular American brands of thread, baking powder, and toothpaste.

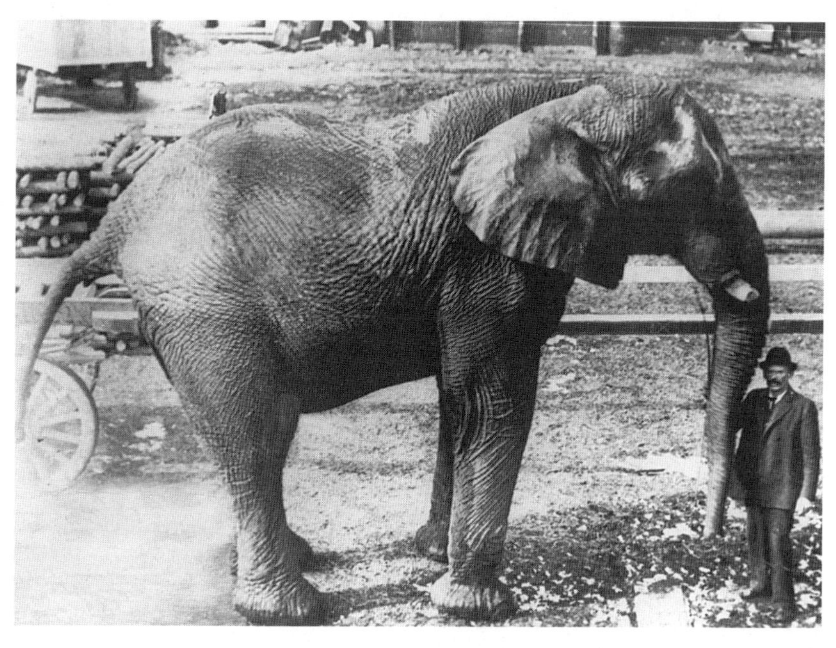

A rare photograph of Jumbo with Matthew Scott, showing the impressive size of the animal. From the private collection of Mr. A. H. Saxon, reproduced with permission.

and leave the rest for Jumbo. One night, however, he himself drank all the beer and then fell asleep. The elephant waited patiently for his nightcap, but finally lifted the sleeping "Scotty" out of his bunk and gently put him on the floor. Matthew Scott, who had known Jumbo for 18 years, at once realized that he had committed a serious faux pas and, with many apologies, handed the elephant a bottle of beer. The bottle was promptly emptied, after which both elephant and keeper went to sleep.

Although Bailey and the circus elephant keepers were quite unimpressed with the gloomy, truculent Matthew Scott, P. T. Barnum was clever enough to observe Scotty's almost hypnotic influence on Jumbo. With his usual foresight and businessmindedness, he signed a long-term contract with Scotty. Jumbo's attacks of fury, which had alarmed Mr. Bartlett and the other attendants at the London Zoo, completely ceased after Jumbo was moved to the United States. This would indeed favor the hypothesis that Jumbo's bad temper was associated with the emergence of his molars. The long-time companionship of Matthew Scott certainly had a calming influence on the elephant. Mr. Scott, through his great stubbornness, managed to launch a scheme of elephant riding at the circus that benefited both Jumbo, who needed his daily exercise, and Scotty's own purse.

P. T. Barnum rightly asserted that Jumbo continued to grow during his sojourn with the Greatest Show on Earth. In 1884, the elephant was more than 12 feet tall and weighed about 7 tons. Once, for a publicity stunt, Barnum arranged for Matthew Scott to lead Jumbo across the newly built Brooklyn Bridge. The bridge withstood this test of its strength, although it vibrated for each of Jumbo's footsteps. Matthew Scott told reporters that he was fearful that it would have collapsed if Jumbo had started to dance. Jumbo became the favorite of the American public, just as he had once been the darling of all Londoners. There were many sentimental pamphlets, books, and poems about Jumbo, in which the great elephant was anthropomorphized to fit the nineteenth-century idea of how an elephant was to behave. The Americans were much impressed by Jumbo's majestic yet gentle demeanor, his impressive memory, and his kindness to young children.

During Jumbo's first 3 years at the circus, the elephant brought in earnings one hundred times in excess of the price for which he had been bought from London. Barnum was well aware that a tour to the British Isles with Jumbo and chosen parts of the circus would prove a very lucrative enterprise. In 1885, he busied himself making plans to cross the Atlantic and restore Jumbo, at least temporarily, to his little friends in London. Jumbo still had a long career in front of him: in captivity, African elephants may well reach the age of 60 or even 70 years.

Jumbo's Death

On September 15, 1885, the Greatest Show on Earth was visiting the city of St. Thomas, Ontario, during its Canadian tour. After the show, when all the other elephants had been loaded into their railway cars, Matthew Scott took Jumbo and his friend, the dwarf elephant Tom Thumb, to their private box cars. They were accompanied by a railway man while trudging along the track toward "Jumbo's Palace Car." Suddenly, a train could be heard rapidly approaching from behind them, and a powerful whistle sounded. The horrified railway man ran toward the train, a long express goods train, hauled by a powerful Grand Trunk Railway locomotive. The train driver probably received the shock of his life when he saw the frantically waving railwayman, and, behind him, Jumbo's immense gray shape looming in the locomotive's searchlight. He threw the engine into reverse and sounded three short blasts on the whistle as a signal to have the brakemen apply the brakes, but the distance was much too short.

As the train approached them, Matthew Scott jumped down the railway embankment and called Jumbo to follow. The elephant was first frightened of the slope, but when suddenly facing the locomotive's powerful headlamp and piercing whistle, Jumbo ran down the embankment, trumpeting loudly. The sheer impetus of Jumbo's 7 tons took him far beyond Matthew Scott, however, and the elephant nearly ran into a high fence. This made Jumbo panic and run back up on the railway track. Matthew Scott desperately tried to drive Jumbo toward an opening between two railway cars. Trumpeting loudly and lifting Scotty, who was hanging on to the elephant by a strap, off the ground with each step, the huge beast ran past them, again from sheer impetus. It is not easy for a 7-ton elephant, running at its utmost speed, suddenly to stop or to turn around. The train first hit the dwarf elephant, which was thrown down the embankment like a football, and then rushed on toward Jumbo, with the brakes shrieking and sparks flying from the wheels.

Just before the collision, Jumbo uttered a tremendous roar; during their 20 years together, Matthew Scott had never heard anything like it. The locomotive struck Jumbo from behind, and the elephant's body was wedged half over, half under one of the railway flatcars. The force of the collision derailed both the locomotive and the coal tender. The driver and the stoker both threw themselves from the locomotive as soon as they realized that a crash was inevitable, and their lives were saved. Jumbo lay on the railway track with severe internal injuries, blood pouring from his mouth and trunk. The skull may have been cracked, but it was certainly not "badly smashed" as claimed by some biographers. In spite of his terrible injuries, Jumbo could still recognize Matthew Scott, and when the keeper spoke to the stricken giant, Jumbo's trunk sought his hand. Clutching Matthew Scott's hand, the elephant died some minutes later. Poor Scotty wept bitterly and could only with difficulty be persuaded to let go of the elephant's corpse.

Before the circus people had presence of mind enough to post a guard on Jumbo's huge body, it was attacked by rapacious souvenir hunters, who cut off hairs from the tail and filed the tusks, like a flock of harpooners pitching into the body of a huge, dying whale. According to one source, there were attempts to cut off both the tail and the trunk as well as to cut chunks of meat out of the flanks. When poor Matthew Scott, who had fallen asleep beside the huge carcass, exhausted by grief, finally awoke from his slumber, he became nearly hysterical when he saw that a huge slice had been cut out of one of Jumbo's ears. The St. Thomas police went out in force to prevent further outrages of this kind, and a policeman was posted to guard the elephant night and day. P. T. Barnum and James Bailey were of course most distraught when they heard that the star of their circus had been killed in a railway accident. They accused the railway of negligence, but the officials of the appositely named Grand Trunk railway asserted that anyone leading their unruly beasts along the railway had to watch out for themselves, since if there were railway tracks on the ground, it was most likely that trains would run on them. There was a story at the time that a young boy, who had been sent to St. Thomas to warn the stationmaster that the extra goods train would pass through the station, had stopped for doughnuts on the way and had arrived too late. If true, this tale does not inspire confidence in the Grand Trunk railway's signaling system.

As befitting one of the world's most famous animals, Jumbo's death gave rise to a great variety of urban myths. Like the death of President John F. Kennedy, many lurid tales and crackpot theories were recounted about Jumbo's last hours in life. According to P. T. Barnum's own version, Jumbo died a hero's death to save his little friend Tom Thumb. Sacrificing his own life, Jumbo had snatched the dwarf elephant with the trunk and pulled it away to safety, moments before being hit by the train. Some writers of popular books on the American circus have claimed that Jumbo died through mere stupidity and that the elephant had charged the train, bellowing with rage. Another, more lurid version stated that Jumbo had been more than usually addicted to his rations of ale and whisky on that fateful day and that the huge beast had reeled toward a drunkard's death in an alcoholic stupor.

Just 2 weeks after Jumbo's death, a remarkable conspiracy theory was advanced by a hack journalist C. F. Ritchel, an old enemy of Barnum's, who was writing in the Hartford *Sunday Globe.* He asserted that Barnum had deliberately planned Jumbo's death, since he knew that the elephant was badly ill with tuberculosis. A dramatic accident would generate a furor of publicity, which would benefit the circus. After reading this perfidious calumny, P. T. Barnum immediately sued the publisher for 50,000 dollars. The journalist considered it to be highly suspicious that Barnum had made a deal with Ward beforehand, but the editor of the *Globe* went to Barnum's headquarters in Bridgeport and obtained testimonies from Matthew Scott and several independent eyewitnesses who had been present when Jumbo died. This proved that Barnum, for once, had been telling the truth. When the irate editor returned, Ritchel was sacked on the spot, and the next issue of the *Globe* contained a mealy-mouthed apology to Barnum.

Another, even more daring conspiracy theory was advanced in the 1960s. Jumbo's incredible flatulence had "stunk up" the entire circus, and out of sheer embarrassment, Barnum had decided that the elephant must go. An unscrupulous English animal keeper turned assassin was instructed to shoot Jumbo with a powerful pistol, while P. T. Barnum himself supervised the elephanticide from a safe distance. Jumbo was shot in the eye and fell on the dwarf elephant, nearly crushing it to death! The circus train had started moving just at that moment to cover the sound of the shot, and to Barnum's chagrin, Jumbo fell between two of the cars, damaging the skull. The cunning showman had arranged for Jumbo's hide to be stuffed in advance, and taxidermists were standing by. All these fanciful tales are nothing but lies and cannot be reconciled with the actual events of that fateful evening in St. Thomas, as testified by several independent eyewitnesses, including Matthew Scott himself. A later reexamination of the elephant's skeleton further implies that Jumbo was most probably hit from the back, that the cause of death was internal injuries caused by the impact of the locomotive, and that the head was wedged between the engine and a railway car with great force. Nor has any trace of projectiles been found on examination of the skeleton and body.

Jumbo is likely to have been the only animal whose death became front-page news, as in the demise of a prominent politician or a royal personage. The news of Jumbo's tragic death was telegraphed all over the world, and many newspapers printed lengthy obituaries of the King of Elephants, often with many sentimental comments fetched from Barnum's untruthful account of Jumbo's hero's death. In an editorial in the *Spectator*, it was emphasized that "It was his destiny to attract more attention than any other elephant which ever existed, and it pursued him till his death. Alone of his race he has been killed by a railway locomotive." In the *Daily Mail*'s obituary, Jumbo was eulogized as "The pillar of a people's hope, the centre of a world's desire." Mr. Punch, with his usual attitude of cheerful, materialistic unconcern, mockingly wrote:

> *Alas, poor Jumbo! Here's the fruit*
> *Of faithless Barnum's greed of gain.*
> *How sad that so well trained a brute*
> *Should owe his exit to a train!*

In Britain, at least one child was baptized Jumbo during the Jumbo-mania of 1882, and it seems as if several American youngsters were named after the famous elephant. A young woman who bore a child weighing

 10½ pounds was widely publicized in the American newspapers after she had insisted that her son was to be named Jumbo. When I was a medical student in the mid-1980s, I met an old Swede who was actually named Jumbo. It was not a nickname, since it occurred both on his birth certificate and in his hospital records. I was not then familiar with the famous elephant's life story, but when I asked him about his strange name, which had already caused great merriment on the ward rounds, the near centenarian replied, with a deep sigh, that he was born on the same day the great American circus elephant had died. His father, a Swedish-American, had seen Jumbo in the United States and been vastly impressed. When he returned to Sweden, the Swedish newspapers were full of the tragic death of Jumbo in St. Thomas. He later found that, on exactly the same day, his wife had given birth to an extraordinarily large child, and the name took little coining. It is unknown, however, exactly how the local rector was persuaded to baptize young Jumbo, but probably a large donation to the church funds made him more appreciative to this novel addition to the Swedish stock of first names. His sister later told me that, throughout his life, this Swedish "Elephant Man" had had to pay dearly for his parents' eccentric whim. Although oversized as a baby, he later became a very short, shrill-voiced boy with an extraordinarily long nose. When he first entered grammar school, the teacher gruffly ordered him to "Remove that false nose this instance!," and when the boy presented himself as Jumbo, the teacher nearly had a fit, believing that his leg had been pulled in front of the entire class. Later, young Jumbo actually tried to become accepted at a police academy. The board of police examiners laughed to the point of falling over when he introduced himself in his peculiar, shrill voice. "Jumbo, you said, little fellow! Ha! Ha! Ha!" shouted one of them, and another replied "Well, at least he has got the *trunk!*"

Jumbo's Widow

Although Jumbo's sudden death was a severe shock to P. T. Barnum, the clever and far-sighted showman had made plans how to handle such an emergency long before it occurred. For several years, he had cooperated with Professor Henry Ward, of the Natural Science Establishment in Rochester, New York, who used to mount and stuff those deceased circus animals that Barnum saw fit to have preserved for future generations. In 1883, Barnum had written to Ward, and a contract had been drawn up that Ward was to stuff Jumbo's skin and mount the skeleton if the famous elephant expired through disease or some accident. It took the professor, his assistants, and six butchers 2 days to dissect the elephant and to prepare the hide and skeleton. A man named Peters was appointed to perform the "inside work"; he likened himself to Jonah inside the whale. At regular intervals, he had to crawl out to get some fresh air, "looking a little white around the gills." Jumbo's stomach was found to contain an amazing collection of objects: numerous coins—gold, silver, and copper, of both British and American coinage; a policeman's whistle; a toy pistol; a bunch of keys; several nails, tack, and buttons; and the brass figure of a pig.

An American wrote to Barnum, desiring to buy Jumbo's eyes, while another wanted to purchase a barrel of elephant's grease, for use against rheumatism. Jumbo's heart was purchased by Cornell University; according to Barnum, it weighed 46 lbs. It was not intended for the provision of free elephant-heart hamburgers for the Cornell students but was instead preserved for posterity at the Veterinary College. When, in 1986, Dr. J. Shoshani tried to trace the heart, the Cornell officials could produce only a huge empty glass jar in which it had, according to the labeling, once been kept. One of the more seasoned veterinarians could recall that he had, 40 years previously, seen Jumbo's heart in this jar. Whatever had happened to it in the meantime is uncertain, but it remains a distinct possibility that the specimen had dried out after the fluid (spirits) in which it had been kept evaporated; this would not have been the first time a "pickled" specimen came to grief in this way. Inquiries at Cornell University, in 1997, did not provide any clues.

Professor Ward and his men toiled for 6 months to mount Jumbo's skeleton and prepare and stuff the enormous hide. The Smithsonian's Spencer Baird pointed out that Jumbo's hide could be stretched during mounting, and the showman delightedly seized on this hint, ordering Ward: "By all means let that show as large as possible. It will be a grand thing to take all advantage possible in this direction! Let him show as a mountain." It was soon apparent to Henry Ward that the businessminded P. T. Barnum had made another clever deal when he drew up the contract to have Jumbo's remains preserved for posterity: his salary of 1,200 dollars was not even enough to pay for the costs of stuffing and mounting materials. Professor Ward wrote to Barnum to ask for more money, but although he gave Mrs. Barnum an inscribed slice of one of Jumbo's tusks, he received none. Instead, P. T. Barnum cashed out on a grand party for politicians and journalists, during which Jumbo's skeleton and stuffed hide were uncovered amid thunderous applause. Small fragments of Jumbo's tusks had been ground and mixed into a

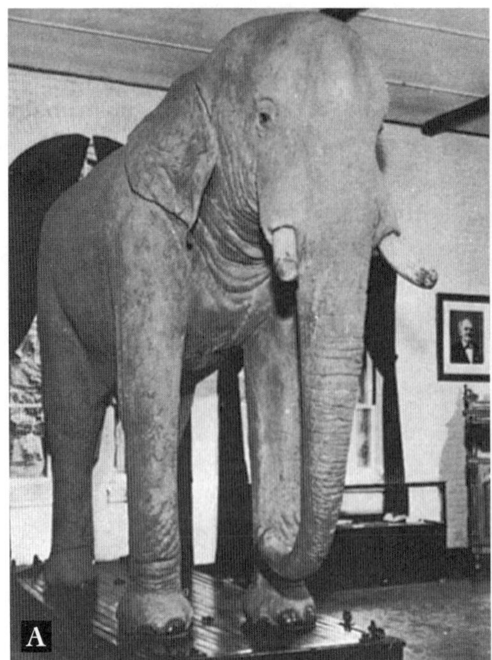

The stuffed Jumbo (*A*) and his mounted skeleton (*B*).

jelly that was consumed by the journalists, politicians, and socialites invited by Barnum. To actually eat parts of the body of the guest of honor was a novel event even at a society banquet! Professor Ward gave a lecture about the preparation of the skeleton and stuffed body, and his skillful work was much admired by the newspaper men.

P. T. Barnum then took both versions of Jumbo on tour with the circus for several years. The famous elephant's power of attraction to the public was nearly as great as during his lifetime. Furthermore, Barnum bought the stolid old elephant cow Alice known as "Jumbo's wife" during their time together in London. In 1886, a grand parade was arranged at Barnum & Bailey's Greatest Show on Earth. While a huge brass band played funereal music, Jumbo's skeleton and stuffed body slowly traveled round the arena, standing on two large carts driven by morticians dressed in deep mourning and pulled by horses with black capes. The grieving widow Alice followed, wearing a widow's cap that was an exact copy of that of Queen Victoria, although of course several sizes larger. Then came the other circus elephants, who had been taught to hold black-edged handkerchiefs in their trunks and to wipe their eyes with them at regular intervals. This orgy in vulgarity, which reputedly was staged by P. T. Barnum himself, was repeated many times and was always greeted with enthusiastic cheering and applause. Late in 1887, the Bridgeport winter quarters of the circus were ravaged by fire. The two Jumbos were saved just in the nick of time, but poor Alice expired in the flames.

The Immortal Jumbo

After a few years, P. T. Barnum became aware that Jumbo was losing his value as a popular attraction; furthermore, the stuffed elephant had gradually become the worse for wear as a result of surreptitious travel on the railways. In his old age, Barnum was a wealthy and generous patron of the arts and sciences: in 1889, he donated Jumbo's skeleton to the American Museum of Natural History in New York, and the stuffed elephant to Tufts College. The latter seat of learning was much supported by Barnum, who was a member of its board of directors. P. T. Barnum had kept the right to take the two Jumbos with him on tour again, however, and later in 1889, both versions of the famous elephant accompanied the Barnum & Bailey circus on its tour to England. At the Olympia just outside central London, the circus was visited by fifteen thousand people each day, many of them, like Queen Victoria and the Prince of Wales, paid their respects to the immortal Jumbo. The innumerable children who had said their prayers every night to save the elephant had probably never expected that their prayers would be fulfilled in this dismal way, and that their beloved Jumbo would be returned to them in duplicate.

Barnum had feared that the Jumbo incident would have made him permanently unpopular with the British nation and that he would be blamed for indirectly causing the elephant's death, but this was not the case. Indeed, P. T. Barnum was more popular than ever, and he was many times applauded in the circus arena and cheered by the crowds in the streets. The only untoward incident occurred when he was viewing a display of the horse guards: from the royal box, the Prince of Wales was glaring at him through a quizzing glass. During a pause in the ceremony, the witty prince sent him a note that he had observed Barnum's enthusiasm for the performance of the horse guards and that he sincerely hoped that the wealthy showman would not purchase them and bring them with him to the United States.

After the British tour, the two Jumbos were returned to the American Museum of Natural History and Tufts College, respectively. Old Matthew Scott accompanied the stuffed elephant to Tufts College, since he seems to have had quite a long-term contract with Barnum; at least, the showman was kind enough to keep paying him a modest salary for keeping the stuffed elephant clean and tidy. With time, the hunched, lonely figure of the old elephantkeeper became increasingly pathetic: throughout the day, he sat by the huge, stuffed elephant, dusting it with a broom and speaking to it as if it was still alive.

P. T. Barnum died in 1891. During his early career in show business, he had been quite a rogue who had imposed on his credulous countrymen with his tall tales of General Washington's 152-year-old nurse and the Feejee Mermaid. In his old age, however, he became a wealthy and generally respected magnate. Circus historians have agreed that he was the greatest American showman of the nineteenth century, who shaped the popular entertainments in that country even more strongly than Walt Disney did the century after. Barnum was also an important popularizer of zoology and natural history in the United States.

For some reason, Jumbo's old keeper Matthew Scott had expected to receive a legacy from Barnum. He was left a fine brass chest, which proved to contain only a signed copy of Barnum's autobiography. This was a severe disappointment to the gloomy little elephant keeper, who became increasingly odd in his old age. In spite of his earlier claims to have supported teetotalism, his habits became increasingly dissipated. The circus continued to employ him as a keeper of various small animals well into the twentieth century. He never returned to his native land but died at the Bridgeport almshouse in 1914.

The stuffed Jumbo remained at the Barnum Museum at Tufts University for many years. The elephant became something of a mascot to this seat of learning: it was depicted on the college flags and on the shorts of the football and athletics teams. According to an old tradition, a coin dropped into Jumbo's trunk just before an examination was said to ensure top grades. Jumbo remained the pride of Tufts University until 1975, when the Barnum Museum was gutted by fire; the flammable stuffed elephant literally went up in smoke. The students mourned the loss bitterly, and it was remarked that college life would never be the same again without Jumbo. Although the fire had destroyed values for more than 2 million dollars, it was the old stuffed elephant that was missed most sorely.

Jumbo's skeleton was exhibited at the American Museum of Natural History for many years, before being taken over to the museum's stores in 1977, where it was kept together with several whale skeletons. In 1993, Jumbo's skeleton reclaimed its position in the open exhibition when the American circus celebrated its 200-year anniversary, and it was seen by millions of spectators, just as in the old days. The lower parts of the skeleton shone as if polished, from having been touched by innumerable spectators. Jumbo's skeleton is today, together with the slices of the tusks that are kept at various museums, the only memorial of what was perhaps the world's most famous animal. It was shown for several months before again being put into storage.

The American people long held sacred the memory of Jumbo—the legendary elephant who had nearly caused a war between Britain and the United States, who could put the trunk in through the upper windows of a five-story building, and who died a hero's death at the St. Thomas railway depot. In 1935, a popular Broadway musical was loosely based on the life story of the famous elephant and was of course eponymously titled *Jumbo*. The musical was later filmed by MGM, with Doris Day in the leading role; although she did not dress in an elephant's costume, the film was nevertheless a great success. In 1985, the bicentennial of Jumbo's death was celebrated in the United States amid much ballyhoo. Jumbo posters, Jumbo matchboxes, and jubilee coins with the elephant's effigy on them were sold. At St. Thomas, a colossal, full-scale statue of Jumbo was erected just by the motorway entrance to this city.

In several languages, the name Jumbo has since been used as a denominator of elephants in general. Since the early twentieth century, the word *jumbo* also has been used to signify an unusually large object within its category. In both Britain and the United States, particularly large garden products were thus called jumbo cucumbers and jumbo pumpkins. This usage has since become widespread in large parts of the globe, and the ghost of the famous elephant has been invoked in the most unexpected combinations. The French philologists have rightly objected to these foreign vulgarisms invading their noble language, but it is still possible to embark "un Jumbo" at the Orly airport. Those Americans who wish to commemorate the famous elephant could have a celebratory dinner in which they might dine on Jumbo prawns served with Jumbo peanuts and Jumbo olives, before the main course of a Jumbo burger (probably made from dead cows rather than elephant meat) served with a Jumbo martini.

There is no doubt that in his time, Jumbo was the largest elephant in captivity. P. T. Barnum never let anyone measure or weigh Jumbo, and in the figures given to newspaper men and zoologists, the elephant's size was often considerably exaggerated. Nor did Barnum consent to Jumbo being photographed in a way that divulged the animal's correct proportions. In the circus drawings, Jumbo was depicted as a monstrously huge caricature of an elephant, who could reach the roof of a 3-story building with the trunk. Barnum even published a print depicting a coach being driven under Jumbo's belly. In the circus advertisements, it was sometimes speculated that Jumbo belonged to a novel species of elephant, or even that he was a mastodon or a mammoth. Strangely enough, the British zoologist Richard Lydekker advanced a similar hypothesis in 1907, suggesting that Jumbo belonged to a novel subspecies of the African elephants, called *Elephas africanus rothschildi* and distinguished by its uncommonly tall stature and the size and shape of its ears. Later zoologists have not agreed with Lydekker's hypothesis, however. The modern opinion is that there are only two subspecies of the African elephant: the bush elephant and the forest elephant. Jumbo is considered a particularly large example of the former subspecies (*Loxodonta africana africana*).

Jumbo's huge size was not the only reason for his astonishing worldwide fame: in modern terms, the great elephant might be called a media product, created by P. T. Barnum's genius for advertisement and publicity. The characteristics attributed to Jumbo bore the stamp of the anthropomorphizing of animals that was common among nineteenth-century journalists. The life of the famous elephant became a fairy tale created by Barnum's craving for publicity and the newspapers' need to sell copies. Jumbo was a worthy forerunner of the present-day rock stars and television pseudocelebrities in that respect. If Jumbo had evaded capture when he encountered the Arab hunters during his stroll near the Settite river on that fateful spring morning in 1861, his life would have been completely different. Instead of embodying the late-nineteenth century idea of how an elephant was to act and behave, he would have lived out the natural existence of his species, until it would have been ended by old age or by the bullet of an ivory-hunting sniper. If he had been allowed to remain in Africa, Jumbo would not have been as large, since an elephant living in the wild rarely reaches the size of one leading an inactive life in a zoological garden with ample provisions of food. He would nevertheless have been a more apt representative of his species than the pampered circus elephant, who was forced to enact many undignified spectacles, both before and after death.

Animals on Trial

ON AUGUST 17, 1487, SEVERAL PEASANTS AND landowners complained before Monseigneur Jean Rohin, the cardinal bishop of Autun, that a multitude of slugs were devastating the young crops in their fields. The cardinal bishop was much angered by this unprovoked invasion of his diocese. He ordered that public processions be held for 3 days in every parish. Three times, the slugs were solemnly requested by all the vicars and curates to leave the diocese of Autun; if they ignored this warning, the cardinal bishop would excommunicate them and smite them with his anathema. Faced with this lethal threat to their future existence and well-being, the slugs rapidly made themselves scarce, and a thanksgiving was held in Autun to celebrate this triumph for the holy church.

Some years later, another danger threatened the diocese of Autun. A host of field rats was laying waste the vineyards and barley crops, and the peasants again turned to the cardinal bishop. This time, prayers and processions did not affect the havoc wrought by these voracious rodents. For their defiance, the field rats were summoned before the ecclesiastical court of Autun by the bishop's vicar general, who was also the court president. A young lawyer named Barthélemy de Chassanée was employed as their defense counsel. He probably realized that he was facing a particularly difficult task when he saw the pompous clerics, dressed in their best habits, elaborate at length about the bad character and notorious guilt of the accused rodents before a vast crowd of peasants dressed in their Sunday finery, who were listening to their tirades with rapt attention.

When Barthélemy de Chassanée was asked why his clients did not attend the trial, he cleverly pointed out that the rats lived dispersed throughout a large part of the diocese and dwelt in many villages. A single summons was not enough to call them all to the cathedral, and there were precedents that people who had not been summoned to appear at a trial in a fully correct manner could not be prosecuted in their absence. The vicar general and the cardinal bishop could not contest this argument, and they had to adjourn the court. Every vicar and curate in the entire diocese was ordered to read the summons to the rats, both from the pulpit and in the fields, while the churchwardens rung their bells until they collapsed from fatigue. At the new trial, Barthélemy de Chassanée again sat alone at the bench of the accused, without being buried under a mountain of squeaking field rats. When he was asked why these disobedient rodents had again failed to appear, Barthélemy de Chassanée replied that they had not been able to reach the cathedral in safety, since their sworn enemies, the cats, had been lurking in the hedges and passages, intent on their lives. Barthélemy de Chassanée argued that there were precedents that a person cited to appear in court, who cannot come with safety, was not to be punished in his absence. The point was argued at some length, but the outcome was again that the court was adjourned, and the rats saved.

The Barrister of the Rats

According to tradition, Barthélemy de Chassanée's skillful defense of the field rats of Autun attracted the interest of his superiors; from that day onward, he enjoyed a meteoric career as a consulting lawyer. Although

Barthélemy de Chassanée was once or twice consulted about animal trials, he preferred having counts, marquesses, and wealthy merchants as his clients, rather than destitute rats and insects, and he amassed a considerable fortune during his long and distinguished career within the French legal system. Barthélemy de Chassanée ended up as president of the *Parlement* of Provence. He was an opponent of the persecution of the Waldensian heretics in Mérindol, and did not give his consent to these unfortunates being summarily arrested and executed. According to a defender of the Waldensians, he ironically said that if even the rats of Autun had been given a fair trial, then why deny that right to these heretics? This defiant statement did not endear Barthélemy de Chassanée to the authorities, and it seems as if one of his old enemies saw an opportunity for settling the score once and for all. He sent Barthélemy de Chassanée a bunch of flowers that had been soaked in a powerful poison; after smelling the beautiful flowers, the Barrister of the Rats fell to the floor, stone dead. His contemporaries were unanimous that he had been poisoned, since he had previously seemed to be in excellent health; even his debunking biographers, Messrs. Bouhier and Pignot, considered it likely that he had been deliberately murdered. These two gentlemen have doubted, however, as have some twentieth-century French historians, that Barthélemy de Chassanée was really involved in the trial against the rats of Autun. The trial is not mentioned in Barthélemy de Chassanée's own writings, an omission that has been considered a damning circumstance. M. Bouhier believed that Barthélemy de Chassanée had instead, under similar circumstances, defended a party of flies, which had destroyed some vineyards near Beaune. In spite of this, many scholars have repeated the tale of the Barrister of the Rats, and it rests on relatively strong authority: the works of the eighteenth-century Judge Auguste de Thou.

One of the many valuable treatises left behind by Barthélemy de Chassanée was his legal *Consilia,* one of which dealt with the formal requirements in animal trials. If the accused beast had spilt blood during its criminal act, it was to be tried by a secular court; otherwise, it was to be punished through the agency of the church. The secular animal trials, against goring bulls and vicious boar pigs, were usually summary processes, ending with the ignominious death of the animal in question. The ecclesiastical animal trials were directed against a large flock of animals or an entire species: rats, insects, or other vermin. They were much more elaborate procedures: the president, or judge, was usually a senior cleric, and the defending and prosecuting council were qualified lawyers. If the animals in question were convicted, they were threatened with excommunication by the clerics involved if they did not promptly leave the territory in question.

Throughout the middle ages, the populace had an immense trust in the bishops and other men of the church; the authoritative clerics were considered almost as demigods, able to conjure demons, to make rain, and to appease hurricanes and thunderstorms. The medieval hagiographers could tell the most astounding tales about the ability of holy men to conquer nature, by means of powerful curses and anathemas. In the late ninth century, Rome was invaded by swarms of locusts. The civic authorities decided to give a reward to any person who could show them a certain number of dead locusts, but this measure did not affect the rapid propagation of the insects. After the secular extermination patrols had thus failed ignominiously, Pope Stephen VI took charge of the campaign against the locusts. He prepared vast quantities of holy water over which a solemn malediction against all kinds of imps, phantoms, and satanic pestilence had been read. After an army of clerics, deacons and choirboys sprinkled the fields and meadows all around Rome with this holy water, the locusts immediately disappeared. Another holy man, St. Agricola, was responsible for the cursing and burning of storks at Avignon in the year 666, a deed that would today have rendered him a stiff prison sentence for gratuitous cruelty to animals and willfully harming an endangered species, if he had not previously been guillotined by members of a militant fraction of the Société pour le Preservation d'Oiseaux. A disgruntled anchorite, known as John the Lamb, once cursed the fishes, which had incurred his anger in some way or other, with results similarly disastrous to the finny tribe.

Slightly more laudable acts were the expulsion of venomous reptiles from the island of Reichenau by St. Perminius in 728 and the appearance of St. Augustine in the sky near Toledo in 1286, where he threw a plague of locusts into the river Targus with the sign of his cross. It is told of St. Bernard that once, when he visited the abbey church of Foigny, both the priests and the worshippers were complaining bitterly about a plague of flies infesting the church; their loud buzzing made the entire building reverberate like a beehive. St. Bernard then excommunicated the offending insects, and the buzzing immediately stopped; the next morning, the grateful clerics had to throw the heaps of dead flies out with shovels. Another authoritative prelate, Bishop Egbert of Trier, was much irritated by having to share his cathedral with numerous colonies of swallows, which were fluttering about and disturbing the services with their chirping. After the birds had committed the additional offense of defiling the Bishop's head, miter and vestments with their droppings when he was officiating at the altar, he solemnly forbade them to enter the cathedral again, on pain of death. It is still a saying in Trier that if a swallow ventures into the cathedral, it immediately drops dead.

If the medieval animals usually fared badly when facing trial, the one brilliant exception is the faithful Dog of Montargis, appositely named Fidèle, which actually appeared in court as the plaintiff in a murder case, according to an old French legend. A certain Chevalier Macaire, an archer in the guards of King Charles V of France, was jealous of an old enemy, Aubry de Mondidier, who had risen highly in the king's favor. When Mondidier was traveling through the forest of Bondy, together with his dog, Macaire shot him dead from an ambush and clandestinely buried the body. The faithful dog reached the royal court, however, and led the soldiers and courtiers to Mondidier's grave. Every time the dog saw Chevalier Macaire, it growled, snapped, and attempted to attack him. This was, of course, considered a highly suspicious circumstance, and although Macaire vehemently denied having anything to do with the death of his colleague, many people accused him of the crime. The king decided that the dog was to be given the right of a plaintiff and that God was to judge the case—that is, that the two were to fight to the death in an arena. If there was any betting on the outcome of this judicial duel—and one rather suspects that there was—the punters were likely to have favored Chevalier Macaire. Whatever his moral shortcomings, the sturdy, muscular mercenary, in his leather jacket, iron gloves, and strong, hobnailed boots, seemed like a formidable enemy to any member of the canine tribe. Furthermore, the distribution of arms in this strange duel was distinctly unfair: Macaire was given a strong, wooden shield with which to ward off the dog's attacks, and a long, heavy cudgel with which to belabor it. The faithful dog was only given a huge barrel, in which it could take cover from the Chevalier's blows. Fidèle eschewed such defeatism, however; when unleashed, the dog went for Macaire's throat like a bullet. The brutal fight that ensued, the dog snapping and growling, would almost have put a match for the heavyweight boxing title to shame. A seventeenth-century drawing of this horrid scene depicts the dog decidedly at a disadvantage: Chevalier Macaire keeps it at bay with the shield, and aims a heavy blow against its head with his cudgel, at the same time violently kicking the animal in an area distinctly "below the belt." A vast crowd of spectators, including many well-dressed gentlemen and elegantly gowned ladies, surrounded the arena to cheer the contestants on. It is recorded that the faithful dog finally gained the

The fight between Chevalier Macaire and the faithful dog, from an old print. Reproduced by permission of M. Jean-Loup Charmet's picture library, Paris.

upper hand in this fight and forced Macaire to cry out for mercy and admit his crime. The victorious Fidèle was honored by the king and his court, and Chevalier Macaire was executed, his body buried in unhallowed soil.

From the Annals of Animal Trials

In the High Middle Ages, the people's faith in the church officials as exterminators of vermin remained very high. But if the curse against the offending animals was to have its full effect, the entire tribe of animals had to be tried and convicted by an ecclesiastical court. Some of these ecclesiastical animal trials occurred already in the thirteenth century, but they rapidly became more and more common in continental Europe during the following

200 years. In 1338, a kind of beetle caused much devastation in the fields near Kaltern in the Tyrol. The beetles were tried, convicted, sentenced, and solemnly cursed in the name of the Holy Trinity. It is recorded that, owing to the sinful lives of the people in these parts, the beetles did not heed this anathema until several years had passed and the people had repented their wicked ways.

In 1479, another famous animal trial took place in the court of the bishop of Lausanne. This time, the inger were on trial for damaging the young crops in the fields. The prosecutor brought forward some cunning arguments to denigrate the defendants. They had not been present on Noah's Ark but had been hovering above it, defying the anathema of God. The prosecutor also claimed that the inger were not animals but rather imperfect, spontaneously generated creatures bred from putrefaction. The defendant of the inger was powerless to prevent that his clients were excommunicated by Monseigneur

The Bishop of Lausanne cursing the inger. From a nineteenth-century Swiss print in the author's collection.

Benedict de Montferrand, Bishop of Lausanne, using a solemn malediction beginning with the words: "Thou accursed, infernal foulness, thou inger, which shalt not even be mentioned among the animals . . . "

In a German trial against a bear in 1499, the defending counsel objected that his client could be sentenced only by a jury of his peers, but the judge did not permit the recruitment of twelve other bears as jurors. The defense had as little success in a trial against the field voles that took place in the Tyrol in 1519, in which the rodents were sentenced to eternal banishment from the territory in question. The only leniency achieved by their counsel was that the expected mass migration of voles should be granted safe conduct from cats and dogs and that, in the name of humanity, pregnant females and their young were to be given 2 extra weeks to prepare their move.

Perhaps the best known of all animal trials was held in 1587. In the still justly famous wine district of St. Julien in France, the young crops in the fields were ravaged by a species of greenish weevil, called *charançon* by the locals. On the thirteenth of April 1587, the insects were summoned before the tribunal of the prince bishop of Mauricenne by the syndics and procurators of St. Julien. An experienced lawyer, Antoine Filliol, was selected as their defense counsel, with a certain Pierre Rembaud as his assistant. The prosecuting counsel, M. Petremand Bertrand, pointed out that God had created the animals before man and that their purpose in life was to serve the Lord of the Creation as loyal, uncomplaining minions. The animals that were willfully disobeying God's directives, such as these gluttonous, criminal weevils, deserved to be severely punished, whether they acted on their own accord or had been seduced by the devil. The barrister of the weevils refuted these arguments in a clever way. He denied that the insects in question had anything to do with the fiend below; instead, they had been sent by God as a scourge for the sinful inhabitants of St. Julien. To punish these innocent messengers would bring down the wrath of the Almighty on the St. Julien peasantry in earnest. Antoine Filliol then quoted the famous lines from Genesis that everything that creepeth on the earth was given the green herbs for meat. If the court denied the weevils the sustenance allotted to them by God, how were these poor insects expected to be fruitful and multiply? That these hapless insects happened to have been created before man did not give this court any authority to sentence and excommunicate them without reasonable cause.

The prosecution side was seriously shaken by these clever arguments. The trial was adjourned several times, and it almost looked as if the weevils had gained the upper hand, but Petremand Bertrand and François Fay, the procurator and advocate of the plaintiffs, thought of a cunning compromise. On June 29, the church bell was rung and all the inhabitants of St. Julien were called to a public meeting just after mass. Facing the crowd of grumbling *vignerons*, François Fay had to admit that the outcome of the trial was very uncertain indeed, and that the weevils had several arguments speaking in their favor. He recommended a compromise: if the court was to banish the weevils from the vineyards, a parcel of land on the village commonage had to be given to these insects, as their private reservation, where no human was to disturb them. After much discussion among the vine growers, a territory called La Grand Feisse was selected as the weevils' future home. It was inspected by the advocate and procurator and declared to be fully adequate for the weevils. The council then drew up a formal agreement of lease, allowing the insects sole rights to all pasture within La Grande Feisse in perpetuity, but reserving for the villagers the rights to pass through this territory, to make use of its springs of water, and to work the mines of ochre and other mineral colors found there. When the trial was resumed on July 24, Petremand Bertrand called everyone's attention to this generous offer, as he called it, but Antoine Filliol cautiously required everything in writing as well as good time for deliberation before he passed his judgment on this novel proposal.

When the legal wrangling was restarted on September 3, Antione Filliol declared that his clients had turned this rascally compromise down, with scorn, since the reservation allotted to them was far too small and barren of vegetation. The weevils demanded a full pardon, with costs, and an ample fee for their own lawyer paid by the court. The point of costs was becoming more and more important as the trial dragged on for month after month, and the fees of everyone involved kept rising. Petremand Bertrand maintained that the insect reservation was amply provided with delicious plants and herbs; the disgruntled vignerons probably uttered the French equivalent of "Hear! Hear!" as he pointed out that the wretched insects could hardly expect to be given an entire vineyard of their own, with a comfortable château to sleep in! To resolve this stalemate, the court president ordered a committee of impartial experts to inspect the intended insect reservation and submit a written report about its suitability. The trial did not come to an end until December 20, 1587, and thus lasted more than 8 months. It is recorded that the cost for clerical works, including seals, was 16 florins, and the fee of the pioneer forensic entomologists who surveyed the reservation 3 florins. The final outcome of this exciting courtroom drama is unknown, although the legal proceedings, recorded on 29 folia, are still kept; the sentence could have been either way. It may be speculated that the weevils were not satisfied with the verdict, however, since the last leaf of the records has been destroyed by insects!

Animals with the Devil in Them

During the heyday of ecclesiastical animal trials, it was commonly believed that Satan could inhabit the bodies of animals and control their actions. Already the Bible had provided a drastic example of this: the Gadarene swine, which were possessed by the devils that Christ had exorcised from an insane man. The entire herd of swine, two thousand heads in all, threw themselves down a high precipice into a lake, where they died. From the protocols of the older animal trials, it is sometimes apparent that the snails, rats, or insects at the bar were considered as the devil's disciples; the ceremonies used to expel them from a certain territory resembled the curses of exorcism of devils from a possessed human being. It was common knowledge that certain animals, such as basilisks and snakes, were particularly allies of Satan. If a snake crawled down the throat of a farmer sleeping in the fields, this unfortunate countryman became possessed by the devil, who reigned absolutely among the pleasant fumes of the wretched man's intestinal tract. Certain theologians believed that fierce and powerful demons could inhabit animals as small as a fly or a louse. As proof, they referred to a startling occurrence from the Erzgebirge in Germany. A thirsty housemaid wanted to drink a pint of beer at an inn one hot summer day. There was a fly in the beer, but she either did not see it or decided not to call the waiter for such a petty matter. The gruesome result was that she swallowed both the beer and the fly and became possessed by the devil inhabiting this insect. This diabolical possession lasted for more than 2 years; she foamed at the mouth and uttered loud oaths and maledictions once a crucifix was held before her. At length, a priest managed to exorcise the fly in a ceremony lasting 12 hours.

A particularly difficult animal trial took place in 1394, at Mortaign in France: a sow stood accused of blasphemy. During its desperate scavenging for food, it had roamed into a church and eaten a wafer left lying on the altar. The priests debated whether this act had transubstantiated the sow's flesh into that of Christ: should it be slaughtered or revered, and should the wafer be honored even inside the pig's stomach? Finally, they decided to execute the pig: there was general relief among the curés when no trace of the wafer was found within the swine's intestinal canal; in that case, they would have had to eat it to show the Savior due respect.

In the sixteenth and seventeenth centuries, it was an oft-debated theological conundrum whether some, or even all, animals were possessed by demons. This problem was discussed by some of the most distinguished scholars of the time and inspired more than one doctoral thesis. In 1739, the eccentric French Jesuit Père Bougeant provided what he considered to be the final solution to this ancient philosophical conundrum. Every time a savage or a heathen was christianized, or an infant baptized, the devil previously inhabiting this individual was simultaneously expelled. Since the Christian religion was making great headway among the American Indians, African savages, and Chinese idolators, this rapidly growing proletariat of unemployed demons was unable to find further human hosts. They were left to haunt the animal creation, giving rise to unpredictable behavior and criminal tendencies in previously docile bulls, dogs, sows, and horses. When other theologians objected, with the argument that they certainly could not consider their faithful dogs and purring, content pussycats as demons from hell, the wily Jesuit responded that they should instead thank God for giving them such pleasant devils to consort with.

These bizarre ideas about possessed, demonic animals were also widely spread among the common people. In fifteenth-century Basle, Prince Radziwill had given the townspeople an uncommon present: a bull moose from Sweden. It was kept in a small enclosure of its own and was regarded with superstitious fear by the common people, who were convinced that it was the devil. Finally, a religious old woman fed the strange animal an apple that had been stuck full of sharp pins and needles. After the moose died as a result of this dastardly attempt on its life, fires were lit on the streets of Basle to celebrate the death of the devil.

Murderous Sows

Throughout the Middle Ages, it was customary in most European cities that pigs were allowed to roam the streets at will. Hundreds, if not thousands, of hungry, filthy, leprous-looking pigs were grubbing about in the sewers and heaps of offal; even the fashionable high streets smelt like a pigsty. It was of infrequent occurrence that anyone actually fed these pigs; they were left to fend for themselves, fighting with the street mongrels for the possession of offal and filth from the gutters and refuse heaps. A desperate, starving sow

A drawing of various criminal animals, including a pig having a meal of child's trotters. From the sixteenth-century legal treatise *Praxis Rerum Criminalium*, by Jodocus Damhouder.

with small piglets could not afford to be fastidious in her choice of victuals; every animal smaller than herself was in constant danger of being attacked. Playing little children were by no means safe from these ravenous swine, and the street pigs not infrequently fell foul of the law.

There are many instances of swine being arrested and executed for having attacked, or even killed, human beings. In Paris, the disadvantage of having hordes of swine roaming the streets had become apparent already in the twelfth century, after a prominent member of the royal family had experienced a strange "traffic accident." When Prince Philippe, a son of King Louis the Gross, came riding along the Rue Saint Jean, along with some of his courtiers, a large boar pig suddenly bolted out from an alleyway. It did not give way to the royal entourage and violently ran between the legs of the prince's horse. All three parties involved fell heavily to the ground; the young prince was precipitated from his horse with such force that he expired shortly after. The sources do not tell what happened to the horse and the boar pig, but if the latter survived the accident, it is unlikely to have been kindly treated by the authorities; perhaps it was punished for dangerous driving by being made into fricassée!

In the annals of animal trials, there are records of more than forty pigs being prosecuted for attacking human beings; the earliest of them from the thirteenth century. In 1386, an infanticidal sow was tried and convicted in Falaise. It had bitten a child in the face and arms, and the barbarian sentence was that it was to be mangled and maimed in the head and forelegs before being garrotted and hanged at the village scaffold. This macabre ceremony was enacted before the Vicomte de Falaise and more than five hundred pious Frenchmen gathered to see the murderous sow pay for her evil deeds. The execution of the sow was immortalized in a fresco in the Church of Holy Trinity in Falaise, which was made at the order of the Vicomte himself, who was probably proud of the part he had played in teaching the criminal elements of the porcine tribe a hard lesson. Large parts of the church was destroyed by the British Army under Henry V in 1417, but the Frenchmen had the fresco redone in the same manner as before. In 1820, this 400-year-old fresco was whitewashed at the order of the church officials when the entire church was repainted and renovated. An engraving of this extraordinary scene, allegedly made when the fresco was still in existence, is reproduced here. It is recorded by Père Langevin, in his 1826 supplement to the *Recherches Historiques sur Falaise,* that the sow was wearing a waistcoat, breeches, and white gloves and that the fresco depicted the hangman and the chief torturer garroting the wretched animal to death. The Vicomte, seated on his horse and wearing a plumed hat, was smugly looking on. From Père Langevin's description, it can be deduced that the mid-nineteenth-century engraving was not, as has been presumed by some historians, a faithful copy of the original fresco but rather a dramatized and much more detailed image of the execution. Some details, such as the position of the wretched swine on the scaffold, and the horsed Vicomte looking on, may well be reminiscent of the original fresco.

In 1403, another murderous sow was executed in Mantes. In the civic archives are still kept an elaborate bill for its upkeep in jail, the cost of cords to bind and hale it, and the hire of a carriage to take it to justice. A separate account was the salary due to the hangman or "Master of High Works," as he was called, who purposely came down from Paris to perform his sorry

Supplice d'une Truie, an engraving by Lhermitte after the original fourteenth-century fresco in Falaise. Reproduced by permission from the British Library.

task. To be able to leave his task with unsullied hands, a pair of new white gloves, value of 2 deniers parisis, were also voted to him. The records of the execution of another pig, which was hanged in Pont de Larche 5 years later, are even more elaborate. It is recorded that the pig's daily board in jail was 2 deniers tournois, equal to that of a human prisoner. It is to be hoped that the jailers did not have the bad taste of serving it a meal of pork chops for its last supper!

On another occasion, a pig that had killed and eaten a small child was convicted not only for murder but also for the sacrilegious crime of eating meat on a Friday. In 1572, an infanticidal pig was actually given a privilege not granted to human prisoners. For times immemorial, the prisoners to be executed at the order of the Lord Abbot of Moyen-Montier had to be naked when mounting the gibbet. The pig was to be led by a rope around its neck and thus could not be considered as completely naked. After due consideration, the Lord Abbot allowed the rope, but it was emphasized that no precedent was to be created and that no two-legged malefactor was to lay claim to clothing on his last journey.

If the owners of these murderous sows could be traced at all, they often assumed no responsibility for the actions of their beasts. Occasionally, they were fined for having let dangerous animals run wild, but more commonly, they were paid compensation by the court, allowing them the sum their executed porker would have fetched if sold on the marketplace. In fifteenth-century Paris, a thoroughbred horse suddenly kicked out and killed a young man. The local court summoned the animal to appear, to be tried for murder, but the horse's owner, who was apparently quite fond of his charger, moved the horse to a stable outside the court's jurisdiction. For this offense, he was severely fined, and the money was used to fabricate the effigy of a horse, which was hung by its neck at the scaffold; the medieval prosecutors of animals certainly did nothing by halves!

Although the pigs were by far the most common animal to appear before the bar in secular courts, many other species occurred in the courtrooms: kicking horses and mules, goring bulls, snappish dogs, and butting billy goats. One or two rams and asses were also executed, as well as a cat that had badly scratched the face of a small child. The reader of Victor Hugo's famous *Nôtre-Dame de Paris,* which takes place in Paris in 1482, will recall that when the unfortunate Esmeralda is charged with murder, her trained goat Djali also faces trial, accused of witchcraft and sorcery. After the goat had demonstrated its ability to count by stamping its hoof, the judge and jury are convinced of its guilt, and it is sentenced to death along with its mistress. Esmeralda is saved by the agile Quasimodo, who swings down in a rope and snatches her from the scaffold. The death-defying hunchback has no opportunity to bring the goat along as well, but the learned animal has intelligence enough to escape on its own accord, to follow Esmeralda through all her miseries for several hundred pages.

Dogs in Jail

Normally, the wretched animals that had killed or severely wounded a human being did not survive their brushes with the law. The most frequent methods of execution were hanging and beheading, but several other, more imaginative manners of death, such as drowning, stoning, burning, and burying alive, were occasionally used by brutal, sadistic individuals. In some rare instances, the accused sow, dog, or bull actually managed to escape its trial alive and well. In 1379, in the French village Saint-Marcel-le-Jeussey, two herds of half-tamed, filthy pigs attacked the son of their swine keeper and tore him to pieces. All the pigs were arrested and jailed. Although only the three large sows had actually been observed to bite the unfortunate boy, their piglets and the other swine in the herds had hastened to the scene of the murder and, by their squealing and aggressive action, clearly demonstrated that they approved of this evil deed. They were regarded as accomplices and were held under penalty of death. While the pigs were awaiting trial, a clergyman, Father Humbert de Poutiers, wrote a petition to Duke Philip of Burgundy, demanding that the piglets and the other swine should be spared. Although he could not deny that they had tried to attack the boy, he pleaded their youth and the fact that they were certain to have been badly brought up by their criminal mothers. The duke lent a gracious ear to this petition, and the pigs were released, to the delight of Father Humbert, who was actually the owner of one of the herds. It is likely that this fact, rather than any humane feelings for the imprisoned porkers, had prompted his actions!

In 1712, a ceremonial procession in an Austrian garrison town was disturbed by the drummer's dog, which suddenly ran forth and bit one of the municipal councilmen in the leg. The drummer was sued for damages, but he denied being the owner of this snappish cur and delivered it into the hands of the court, to do what they pleased with. The dog's punishment was a year's imprisonment in the so-called *Narrenkötterlein,* a pillory or iron cage erected in the marketplace. Drunkards, blasphemers, and petty criminals were thrown into this cage for a few days, to be mocked and jeered by the populace while sobering up. The dog was also accompanied by the village idiot, who was a permanent resident in this cage. An even stranger case comes from late-seventeenth-century Russia. A child in a wealthy family had been butted down a flight of stairs by a tame—or not so tame—billy goat. The goat was sentenced to 1 year in a prison camp in Siberia; its legal owner had to pay the crown 1 kopek a day for its upkeep there.

Some early students of the history of animal trials have found it odd that animals, but not vegetables or dead objects, were put on trial. But the macabre truth is that dead objects have many times faced trial during the history of humankind. The old chronicler Pausanias tells that an enemy of the famous athlete Theagenes, who had once been mauled by him in a boxing match, decided to take out his revenge postmortem. Every year, on the day Theagenes had died, he traveled to the island of Thanos, where a statue of Theagenes had been erected, to flog the statue of his old enemy with a strong bullwhip. These outrages continued for several years, until the long-minded old Greek whipped the statue with such fury that it toppled over and crushed him to death. The statue was then arrested and tried by the Thanos high court. It was found guilty of manslaughter and was punished by being thrown into the sea from a ship. Some years later, Thanos was struck by a failure of the crops and a general famine. The island council decided to consult the Delphic oracle to determine the cause of these bad times, and she advised them to recall all individuals who had been banished from the island. The famine went on in spite of this measure, however. When the councilmen went back to the oracle to complain, she replied that they had forgotten their great countryman Theagenes. The statue was then pardoned, dragged out of the sea, and reinstated on its pedestal. The famine ceased, miraculously, after this belated act of justice had been performed.

Pope Stephen VI, who had saved Rome from the locusts, became notorious by prosecuting the corpse of his predecessor. The deceased Pope Formosus was accused of having usurped his title. His corpse, which had been reposing in its sarcophagus for 8 months, was hauled up from the tomb, dressed in ceremonial robes, and placed in the papal chair of St. Peter's Cathedral. As soon as Formosus had been found guilty, the corpse was pushed out of the chair, and three fingers from the right hand were cut off. The corpse was dragged out from the cathedral, feet first, to be thrown into the Tibre. Stephen VI did not have long to rejoice for having settled the account with his predecessor in such a thorough manner; within a couple of months, he was himself deposed, jailed, and poisoned by powerful enemies. Whatever remained of poor Formosus was resurrected from the Tibre and reinstated into its proper resting place—with all due ceremony.

When Prince Dimitri of Russia, a son of Czar Ivan IV, was murdered in 1591, the insurgents rang the great church bell in the town of Uglich to incite the populace against the authorities. When the rebellion had been quenched, the furious czar dealt summarily with the individuals accused of having started it; the church bell was also at the dock, charged with a serious political offense. It was exiled to Siberia, where it was hung in another church and performed its former office for many years. It was not until 1892 that the bell was pardoned and returned to its former home in Uglich.

In ancient China, it also sometimes occurred that corpses faced trial for the crimes the individual had committed during life. As late as 1888, the corpse of a salt smuggler was brought before the municipal court of Shanghai. He was duly convicted, and the semi-putrid body was decapitated on the city scaffold. At about the same time, even more bizarre things were afoot in the city of Fouchow. A general who belonged to one of China's most illustrious families had fallen in battle. His mourning relatives knew well that, just before he went off to war, he had spent a lengthy session consulting the wooden statues of his favorite gods in the city temple. As the general had failed to return alive, his relatives now prosecuted the gods for having given him bad advice. The viceroy's court found them all guilty as charged, with no attenuating circumstances. All fifteen wooden deities were unceremoniously evicted from the temple, and their heads were chopped off before a large and rowdy crowd. The remains were thrown in a pond, amid cheers and applause from the mob.

Excommunicated Sparrows and Law-Abiding Rats

Already in the Middle Ages, certain humane philosophers and theologians were objecting to the frequent animal trials in France and Switzerland. One of them was the French judge Philippe de Beaumanoir, who considered the criminal prosecution of animals meaningless and barbaric. He was supported by no less authority than Thomas Aquinas, who doubted, in his *Summa Theologiae,* that God would sanction the excommunication of animals. Not even he was able to convince his fellow clerics, however, and the ecclesiastical animal trials instead became more and more frequent. As before, a considerable proportion—in the sixteenth century, not fewer than 19 out of 28 ecclesiastical animal trials—took place in France.

After the Reformation, the old catholic tradition of excommunicating animals was criticized by the strict Lutherans, who considered it superstitious nonsense to put field mice and weevils on trial, and to curse them in the name of the Holy Trinity. There is at least one example, however, of a Protestant parson utilizing these time-honored methods to rid his church of undesirable animal pests. In 1559, the Saxon vicar Daniel Greysser put the sparrows that infested his church under the ban, on account of their ceaseless chattering, their uncleanly habit of depositing their droppings on the heads and frocks of the devout churchgoers underneath their perches, and their scandalous unchastity, which was the subject of much comment. Some disloyal parishioners reported on Parson Greysser for his unconventional methods of keeping down the sparrows, which was an apostasy, they claimed, into the old popish superstitions. Duke Augustus of Saxony, who personally judged this clerical feud, instead commended the ornithophobic parson for his diligence. Daniel Greysser had not, like the bishop of Trier several hundred years before, relied on supernatural means to rid his church of these winged marauders. He had employed several hunters to decimate the sparrows with their bows and arrows and, with his own hands, had smeared the rafters of the church with a viscous bird glue, which was strong enough to immobilize the wretched birds forever, as soon as they settled on these sticky perches.

Even among Catholic theologians, opinion about animal trials changed in the early seventeenth century; many of them adhered to Thomas Aquinas's opinion that God was unlikely to sanction the expulsion from the church of creatures belonging to the brute creation. Already in 1534, animal trials were prohibited in Portugal, and several French dioceses followed suit. In 1717, the pope himself forbade the excommunication of animals, and very few ecclesiastical animal trials occurred after this year. The vast majority of them had been held in France, with some spreading to Switzerland, Italy, and Germany. Few occurred in the rest of the world. In Scandinavia, there is no record of any sixteenth- or seventeenth-century animal trials. The annals of the court of Stockholm provide evidence of a refreshingly down-to-earth attitude toward miscreant livestock. When a cow had gored out the eye of a child, the judge made the salomonic decision that the child's mother was to be given the cow, as a fine to its careless owner. In Britain, a similar attitude seems to have prevailed, and few instances of animal trials exist on record. One of them, the execution of a Scottish dog in the seventeenth century, was one in a series of witch trials, and the dog was suspected to be possessed by one of the witches.

The influence of the Catholic Church is well illustrated by a ludicrous early-eighteenth-century animal trial from Brazil. The Franciscan friars in the cloister of St. Anthony, in the province of Riedade no Maranhão, had been greatly tormented by termites, which devoured their food and undermined their houses and furniture. In 1734, an application was made to the bishop of these parts, who was apparently ignorant of the papal bull issued

17 years earlier, since he was able and willing to summon the termites before his ecclesiastical court to give an account of their conduct. Their defending counsel appears to have taken his bizarre task quite seriously and defended the insects with both pathos and eloquence. Not content with merely presenting the usual argument that the termites were God's creatures and thus entitled to reasonable sustenance, he praised the diligence and industry of his clients, claiming that the white ants were far more laudable in this respect than their opponents in the law court, the Gray Friars. He also, quite reasonably, reminded the court that termites had lived in Brazil long before these Monkish invaders had chosen to settle there. With such powerful arguments in favor of the termites, the monks stood no chance of winning their case. The trial ended in a compromise: the monks had to provide the termites with a reservation, where they could live in peace without interference from the Gray Friars. The court solemnly ordered the termites to leave the monastery and to remain henceforth within their reservation. As the judge read this proclamation, in front of the termite hills, the monkish chronicler reports that all the insects marched out, in strict columns, toward their new abode.

In the countries where ecclesiastical animal trials had played a prominent part, the people were most unwilling to forsake this time-honored method of ridding their fields of gluttonous rodents and noxious insects. In Switzerland, the scepter of St. Magnus was considered to have remarkable vermifugal properties. Between 1685 and 1770, it was regularly carried in processions between various Swiss and German towns. In rural France, curses and processions were used to drive away rats and verminous insects well into the nineteenth century. In difficult cases, the peasants sometimes summoned the animals before the local magistrate. In 1826, a swarm of locusts was convicted of vandalism by the village court of Clermont-Ferrand. Strangely enough, these rural animal trials also occurred in protestant Denmark, which completely lacked a prior tradition of excommunication of animals. In 1711, the village of Als in Jutland was much tormented by a plague of rats, which ate the crops in the fields and dug out the earth with their multiple holes. The village quack advised that the rats be put on trial, and the local counsel sentenced all "rats, field-rats, water-rats and mice, none expected" to leave the village of Als within 8 days. Although they were not observed to walk away in strict marching columns, like the law-abiding Brazilian termites, the complacent Danish rats meekly obeyed this summons. After having been tormented by these ravenous rodents for 3 long years, the entire territory became free of rats. This success paved the way for further Danish trials against rats, held in Viborg in 1736 and in Lyø as late as 1805.

Thieving Dogs and Reactionary Parrots

In early-seventeenth-century France, secular animal trials were still a reality: pigs, bulls, and mules regularly appeared before the bar. In his famous comedy *Les Plaideurs,* Jean Racine made fun of these strange medieval legal relics. A dog is on trial, accused of having stolen a capon. After the defending and prosecuting counsel have presented their verbose and eloquent arguments pro and con, the judge sentences the dog to lifetime hard labor on the galleys, presumably as a ship's dog. The defending barrister then brings in a litter of puppies, the "children" of the accused, and sentimentally pleads with his honor not to make these tender, guiltless creatures into orphans. The judge, who is a father himself, is deeply touched by the barrister's arguments, and the thieving dog is acquitted and reunited with its puppies. The joke about the dog on trial seems rather pointless today, but Racine and his intellectual contemporaries wanted to poke fun at the animal trials, which they found increasingly ridiculous, and a barbarous relic from the Middle Ages.

Toward the end of the seventeenth century, several other French lawyers and intellectuals objected to the animal trials, and the authorities had to relent: fewer and fewer animals faced trial. A contributing factor was that, in the large seventeenth-century cities, the fierce, scavenging street pigs were kept at bay by the city magistrates in a much more efficient manner. In some cities, such as Naples and Grenoble, the city watchmen had the right to shoot them on sight.

In 1792, at the height of the Reign of Terror, an eccentric old invalid named Saint-Prix was arrested by the French police. He was known to have royalist sympathies, and had been accused by his neighbor, the revolutionary constable M. Jardy. Later the same evening, the latter individual wanted to search the house of Saint-Prix, under the pretense that he suspected that certain forbidden books and pamphlets were kept there. As he opened the door of one of the wardrobes, the prying constable received the shock of his life: a large mastiff, dressed in blue livery, leapt out and knocked him to the ground. After a furious fight, the mangled sansculotte, whose trousers had been entirely torn off, was able to leap out through a window. M. Jardy ran to the city hall to make a complaint against the dog. The mastiff was later arrested by a troop of soldiers and was jailed and accused of assault and contrarevolutionary activities. As Saint-Prix faced the revolutionary tribunal, one of the strongest points of evidence against him was that the dog's blue uniform had been made to resemble those of the royal guards. The sentence was that both the dog and its master were to be guillotined. The order for this

The puppies are reprieved: one of the illustrations to an early edition of Racine's *Les Plaideurs*. From the author's collection.

macabre execution, carrying the name of the notorious revolutionary magistrate Fouquier-Tinville and signed by his secretaries, Messrs. Lavillette and Chavet, has been reproduced by M. Jean Vartier in his excellent monograph on animal trials.

The same year, a parrot faced trial in Paris, also accused of counter-revolutionary activities. In a busy street, it had loudly called out, in front of a hostile crowd of sansculottes, "Vive le Roi, Vive nos prêtres! Vive les nobles!" The parrot and its owners, two noble ladies, were arrested and brought before the revolutionary tribunal. In the courtroom, the magistrates tried to persuade the bird to speak, but in spite of their dire threats, it merely whistled disdainfully. The owners of this tactless, indiscreet bird, Madame Louise de la Fiefville and Mademoiselle Françoise de Béthune, were probably most grateful that their parrot kept its beak shut for a change, as they still had to face several difficult questions about who had taught it to repeat such treacherous utterances. They were threatened with the guillotine but were later pardoned owing to a lack of evidence. The reactionary parrot was kept in custody, however, since it had been sentenced to political reindoctrination in the hands of Citoyenne Le Bon, one of the vile *tricoteuses* who delighted to sit near the Paris scaffold and mock its hapless victims as they were led up to be guillotined. In her company, the parrot soon learned to shout "Vive le Nation" as well as a large repertoire of curses and bawdy songs, which is likely to have dismayed and shocked the two noble ladies when the bird was later returned into their care.

Murderous Pigs and Learned Philosophers

The criminal prosecution and capital punishment of animals is a dark chapter in the history of humankind. These macabre and sometimes well-nigh incredible trials have been almost completely forgotten: there is nothing written about them in the schoolbooks or in works on popular history. The majority of legal historians seem to have considered them a puzzling but unfruitful sidetrack in the development of legal thought. In particular, the scholars were at a loss to explain why, at a time when most European philosophers considered animals as little more than automatons, devoid of both sense and reason, these same animals were seemingly put on a level with human beings when facing justice.

The erudite German legal historian Professor Karl von Amira was the earliest scholar to publish a systematic study of animal trials. He pointed out the crucial difference between the complicated ecclesiastical animal trials—

which went on for months, with much eloquent pleading from the defending and prosecuting council—and the summary punishment of murderous sows and bulls through the secular courts. It is true that these two types of trials were different in form and development, but they were conceptually quite similar. At some of the ecclesiastical animal trials, the priests gathered representatives of the prosecuted animals; a bowl of insects or a cage full of squeaking mice was put on the defendant's bench. After the animals had been convicted, they were ceremonially beaten to death, and it was often lamented that their entire tribe had not appeared at the bar. Both types of trials also had similar temporal and geographical distributions. Their centers were in France, from whence they spread to Switzerland, Italy, Germany, and the Low Countries.

In 1937, the learned German legal historian Dr. Hans Berkenhoff added a third category of animal trial. In addition to *Thierstrafe* (secular animal trials) and *Thierbannung* (ecclesiastical animal trials), there was also *Rechtsrituelle Thiertötung,* the ceremonial killing of animals involved in bestiality cases. This distasteful version of animal trials flourished from the fourteenth century well into the eighteenth century. It was of regular occurrence all over Europe that sex-starved peasants were caught in the act of buggering their farm animals. The moral outrage caused by these bestiality offenses was such that both the perpetrator and the animal had to be put to death. The killing of the animal was at least partly motivated by the fear that it might, with time, give birth to loathsome monsters—part human and part animal. Illustrations of such presumed animal-human hybrids are reproduced in several of the old annals of monstrous births. The methods of execution used in bestiality cases were even more cruel and barbarian than in the usual animal trials: the man and the brute he had assaulted were tied to a pole and burned alive on the pyre, or buried alive together in an underground recess.

At times, when some unfastidious Don Juan had been at large, an entire menagerie had to be disposed of: cows, horses, asses, sheep, dogs, cats, and ducks. In his *Magnali Christi Americana,* the Reverend Cotton Mather recorded that on June 6, 1662, at New Haven, Connecticut, there was a 60-year-old man named Potter who had "lived in the most infamous Buggeries for no less than fifty years together." As he stood at the gallows, a cow, two heifers, three sheep, and two sows "with all of which he had committed his brutalities" were put to death before his eyes. A particularly macabre side effect of these laws against bestiality was that in France, sexual intercourse with a Jewish or Turkish woman was likened to defiling oneself with a brute beast, since these heathens were little more than animals. A certain Jean Alard, who had kept a Jewish mistress at his house in Paris, was convicted of sodomy and was burned alive, since coition with a Jewess was likened, by the learned court, to copulation with a dog.

Some early students of the criminal prosecution of animals wanted to explain the old animal trials as the consequences of a primitive, superstitious popular culture. The execution of brute beasts was likened to a child breaking a flowerpot as a punishment for having tripped over it. This attempt at explanation has justly been discredited by later historians. There were few instances of criminal prosecution of animals in the early middle ages, but in the fifteenth and sixteenth centuries, a time of both legal and cultural progress, the animal trials became increasingly frequent. Furthermore, the tribunals dealing with the erring beasts were hardly any primitive village assizes: dogs, pigs, and mules were convicted by the *Parlement* of Paris, and in 1621, the Leipzig law faculty gave an opinion about the proper mode of execution of a cow. Similarly, the foremost ecclesiastical animal trials were led by senior clerics.

Another early hypothesis was that the prosecuted animals were anthropomorphized — that they were, in legal terms, treated as human beings. The arguments were that the methods of execution, the costs of the upkeep of the animals in jail, and the bills of the hangman did not differ much between animal and man; the rest of the reasoning, however, does not fit the existing evidence. It is true that the infanticidal sow at Falaise had been dressed up in a man's clothing before the execution, but it was unique in the annals of animal trials in being granted this dubious honor. Interestingly, the original description of the church painting made to commemorate the pig's execution mentions that not only had several hundred people gathered to witness its punishment, but *many herds of swine* had been driven together in the square before the scaffold. The Vicomte de Falaise probably wanted to demonstrate, in a didactic manner, what happened to murderers, whether they were two or four legged.

Yet another opinion was advocated by the eccentric American scholar Edward Payson Evans, who spent 44 years of diligent labor in the Royal Library of Munich, where he studied animal trials and other arcane subjects. His magnum opus, *The Criminal Prosecution and Capital Punishment of Animals,* was published in 1906. Evans started work on this book in the early 1880s and published some abstracts from it as early as 1884; during the years, he had dug up many astonishing instances of animal trials from various French and German archives and periodicals. I am, like all later writers on animal trials, much indebted to Edward Payson Evans for much of the older material on this subject. The book is as witty as it is erudite, and if Evans had not reprinted some of the original documents as an appendix, the reviewers may have believed that it was just a practical joke. Unfortunately, although Edward Payson Evans had had more than 25 years to arrange his material, it appears rather haphazard to a modern reader; like an absent-minded professor, he piled up his cases in an anecdotal manner. The book lacks a coherent summary and interpretation, and a lengthy section on "medieval vs. modern penology" is wholly out of place, dealing more with Evans's pet theories on criminology than with animal trials. He was a staunch advocate of the "hemp cure" for dangerous criminals, and advised against mollycoddling with those rapists and murderers who claimed to be mentally deranged. His views on other subjects, such as craniology, the brain weights of famous people, the Dreyfus case, and the theories of Lombroso, are equally forthright and eccentric.

In his analysis of the reasons for animal trials, Edward Payson Evans pointed out the crucial role of the Catholic Church, and the widespread notion that animals could be possessed by demons. It was important for the church to keep up the ecclesiastical animal trials: they united the parishioners and, if the excommunication of the animals was a success, inspired confidence in the omnipotence of the church. The people were kept busy praying and walking around in processions in the fields. The clerics did not neglect this welcome opportunity to reclaim unpaid tithes and to exhort promises of chastity, sobriety, and the banning of playing cards and gambling with dice. If the conjuration of the animals was a failure, the clergymen used to blame the sins of the people to avoid losing face. One would, under these circumstances, have expected the people to murmur against these ineffective exterminators of vermin who had forced them to pay money, walk in processions, and abstain from wine and red meat, but who nevertheless could not show any result of their endeavors.

In a lengthy study of animal trials published in 1981, the American philosopher and Orientalist J. J. Finkelstein roundly criticized Evans's book, although he grudgingly admitted that the vast majority of the raw material that his eccentric countryman had dug out of various dusty archives and forgotten periodicals was beyond reproach. Finkelstein proposed that the medieval interpretation of a passage in Exodus, in which it is deemed necessary to kill the ox that gores a man or woman to death, was the instrumental motive of the secular European animal trials. Such was the horror inspired by the animal killing a human being: that the animal was considered unclean even after having been executed. The Bible not only prescribed the stoning of a goring ox—or, in medieval interpretation, the hanging of a pig that had killed a child—but it also forbade the consumption of the animal's meat. Finkelstein's theory is contradicted by some of the original evidence, however. Sometimes, the Dutch or French peasants wholly disregarded this old superstition, cut down the well-hung pig's body, and roasted the criminal whole. Another serious objection to Finkelstein's theory is that the ancient Greek and Roman legal systems had no tradition of excommunicating or punishing animals. The Romans viewed the animal as a senseless thing, that was meaningless to punish. In cases of animals killing or injuring human beings, they advocated the principle of *noxae deditio,* in which the injured party was given the offending animal as damages. It is hardly consistent with the modern notions of psychology to hand over to the grieving mother a filthy, blood-soaked pig that has just killed and devoured her little child. But it may be argued that this principle remains less barbarous than the torture and execution of a pig dressed in man's clothing, in front of a cheering crowd and a huge herd of other pigs. Finkelstein's theory cannot explain a series of animal trials, begun in the thirteenth century and continuing for more than 500 years, purely on the strength of this Biblical pretext.

According to the medieval, anthropocentric conception of the world, man was created as the image of God and was to rule the earth. This gave man an extraordinary power over the animals, and his laws were valid not only for other human beings but for all creation. The animals had been created to provide the humans with labor, food, and clothes; they were not expected to take any initiatives of their own, least of all to injure any one. A bull that gored its owner to death set aside God's ordinance, just as a flock of locusts that devoured the crops in the fields. Man was not allowed to abuse his immense power over nature, however, and every creature was entitled to a fair trial. The criminal prosecution of animals, which otherwise had no rational raison d'être, upheld the dogma of the church's power over nature, and the illusion of a fair and just society.

Another argument to explain the remarkable longevity of the ecclesiastical animal trials is that they often seem to have achieved their purpose. In many instances, it is recorded that the animals that had been tried, sentenced, and excommunicated gradually became more scarce within a year or two. This is likely to have depended less on supernatural reasons than on the laws of population dynamics of insects or rodents. If the voles, for example, had been multiplying rapidly for 2 years, they were faced with an increasing number of predators and much harder competition for food; a cycle of abundance was followed by one of scarcity. The clerical exterminators of vermin thus had the odds set in their favor. Some of the more dramatic results of the excommunication of insects may also have been due to natural cases; the mass death of flying insects, for example, might have been caused by a sudden frost.

The Legacy of the Animal Trials

In the late nineteenth century, the old ecclesiastical animal trials were gradually forgotten, although some French ethnologists have found traces of them in certain quaint rural ceremonies to "conjure away" insects and vermin. The trials against individual animals have continued well into our own time. Although the murderous pigs of medieval times no longer roam the streets, animals still regularly appear at the bar for having killed or seriously injured people. The last time a pig was put on trial was in Pleternica, Slavonia, in the year 1864. A sow that had torn both ears off a little girl was sentenced to death. The fine paid by the sow's owner was to be used as a dowry for the poor girl, so that the loss of her ears did not prevent her from getting married.

The vast majority of pigs, oxen, and other traditional farm animals today eke out their miserable existences in the torture chambers built for them by humankind; without trial, they are imprisoned for life in dismal "animal factories" far from the human race, who make the acquaintance with the inmates of these disgraceful institutions only in the form of tastefully packaged parcels of meat on the supermarket shelves. Since the goring bulls and murderous sows are today thus incarcerated before even having a chance of committing any crime, the dog has taken over the pig's role as man's worst enemy in the courtroom. In 1905, a traveling salesman was attacked by robbers in Délémont, Switzerland, the two footpads had with them a large, fierce cur, which they set on the wretched man. When the three criminals faced trial, the two men were sentenced to life in jail, while the dog, which had actually mangled the salesman to death, was executed.

The large numbers of pet dogs kept, and the irresponsibility of some of their owners, has forced the regular appearance in court of members of the canine tribe. In most European countries, the owner of a dog is responsible for the actions of his beast. If the dog bites people, it is muzzled and the owner fined; if the owner defies this decision, the dog is put down and the owner even more heavily fined. Jurisdiction in Africa and South America seems to leave much to the individual judges in cases of animals appearing before the bar. In 1991, an Argentinian dog was sentenced to lifetime imprisonment for having killed a 3-year-old child. In 1974, a dog in Libya was tried and convicted of biting a human and was sentenced to a month in prison on bread and water. Even more bizarre, a goat in Tanzania, which had grazed on a private lawn, was sentenced to 4 days in jail by an eccentric judge. Another criminal goat, which had stolen fruit from a street pedlar, escaped with 2 days behind bars. Certain American states have prosecuted dangerous dogs well into the twentieth century. In January 1926, a stray German shepherd in Kentucky was charged with the attempted murder of a small child; it was sentenced to death and executed in the electric chair.

As late as 1960, the 5-year-old Alsatian named Duke was prosecuted for having attacked several citizens of New Canaan, Connecticut. For well-nigh a year, it had surreptitiously chased and bitten passing bicyclists and motorcyclists. An old man riding a three-wheeled motorcycle had been badly bitten in both buttocks, and this incident landed Duke at the dock. Duke's owner, a wealthy lady, claimed that her Alsatian was a nice dog, that only claimed excessive attention from people. Facing the threat that Duke would be shot, she employed a clever lawyer. This gentleman seems to have been as smooth an operator as Barthélemy de Chassanée in Autun 450 years earlier. The vicious cur was not to appear in court; instead, film slides of it as a cute little puppy were shown to the jurors, and Duke's mistress told sentimental tales about her doggy's great devotion to her. The jurors were not unaffected by the counsel's arguments and, to the chagrin of the dog-bitten citizens of New Canaan, Duke's life was spared. The verdict was that the vicious Alsatian was to be imprisoned at a dog's kennel near the city in question. Egged on by Duke's owner, the lawyer appealed even this relatively mild sentence. He was able to demonstrate an affidavit from the animal psychologist Dr. John Behan, the principal of the West Redding Canine College, who had treated Duke for several weeks. Dr. Behan certified that Duke had learned to handle his aggressions and that his personality and temperament had improved greatly during his stay at the canine academy. Judge John T. Dwyer then decided that Duke was to be released into the custody of his original owner. This bizarre trial attracted a good deal of newspaper publicity. When the farmer who had been bitten by Duke was interviewed, he claimed that, in the past, there had been better ways of handling vicious animals—this probably more due to reactionary opinions than to any knowledge of Europe's old animal trials. If he had been consulted instead of the "dog shrink," Duke would have been taken for a final "walk" with his leash chained to the rear bumper of a Chevrolet truck.

The Riddle of the Basilisk

> *Like as the Basiliske, of Serpents seede,*
> *From powerful eyes close venim doth convay*
> *Into the lookers hart, and killeth far away.*
>
> Edmund Spenser, *The Faerie Queene*

IF ANY PRESENT-DAY SCIENTIST WOULD ASPIRE TO write a review of state-of-the-art knowledge in every branch of natural science — medicine, zoology, botany, geology, and ethnography, among others — he would have been called a fool or an incurable enthusiast. Owing to the extreme subspecialization of modern science and the immense progress in molecular biology and experimental medicine, even a dedicated team of researchers, with a limitless supply of empty CD-ROMs to cram full of information, would fail in the attempt. Science Citation Index, the major bibliographical update of each year's progress in medicine, technology, and the natural sciences, is now presented mainly on computer disc due to the torrent of novel information flowing through the ever-growing number of scholarly journals.

It is a matter of debate who was the last person who managed to keep abreast of the advances in all branches of natural science. Many of the old cosmographers, such as Ulysses Aldrovandi, Sebastian Münster, and Athanasius Kircher, were men of immense learning, and their works were far reaching and extensive, although much relying on earlier authorities. Already the earliest modern biologists—Carl Linnaeus, John Hunter, Albrecht von Haller, George Louis de Buffon, and Johann Blumenbach—did not write with such a large scope in mind.

It can be determined with some certainty, however, who was the first to produce such a monumental work. The Roman savant Gaius Plinius Secundus enjoyed a varied career, in turns being employed as an officer, a lawyer, and a district governor. He was also, throughout his life, a copious writer of history and biography. When attempting another project, an encyclopedia of the whole of nature, he proudly announced that this design was a novel one: no Greek or Roman writer had previously produced a work of these extensive contents. In the thirty-seven books of his *Natural History,* Pliny collected the obtainable information from the existing works on all aspects of biology, medicine, geography, and ethnology. A life of diligent study prepared him for this magnum opus. Pliny spent all day in his study, reading; he slept as shortly as possible, and when he ate, bathed, or traveled in a sedan chair, his secretary always read aloud. Like many noble and erudite Romans, he was also a man of action: he spent considerable time as an officer in the Roman cavalry and later commanded the Roman fleet in the Mediterranean. At the eruption of Mount Vesuvius in A.D. 79, Pliny led the mission to evacuate the inhabitants of Herculaneum by sea, with considerable success. But when Pliny himself went ashore, climbing a mountain to study the eruption of the volcano at close range, he was suffocated to death by the sulphurous fumes—a martyr to science if there ever was one. Much relying on earlier works, and abounding with imaginative stories and unsubstantiated theories, Pliny's *Natural History* is today an easy target for pedantic and self-righteous scholars. It should be kept in mind, however, that it was the first work of its kind, incorporating the information in many earlier books on natural history, more than 460 of which were mentioned in the text. The majority of these earlier works are now lost owing to the ravages of time, whereas Pliny's *Natural History* is likely to live on forever.

In the eighth book of the *Natural History,* Pliny describes the strange beast called Catoplebas: it was a kind of largish gnu living in Ethiopia. Pliny wrote that its head was so heavy that its head constantly hung down toward the ground; this was fortunate, since all living creatures that met its gaze expired instantaneously. Pliny then pointed out a fact probably well known to many of his readers: that the basilisk serpent had the same power. This venomous serpent was a native of northern Africa. It was not more than 12 inches long and was adorned with a bright white marking on the head, like a diadem. Pliny wrote that the basilisk does not grovel in the dust like other serpents, but advances lofty and upright. Its sibilant hiss puts all other snakes to flight, and it kills the desert bushes and shrubs by its mere presence. The mere presence of a basilisk scorches the grass and bursts the rocks asunder; such is the poison and evil contained within this loathsome beast. Its Greek name was Basiliscus, a diminutive form of *basileus,* meaning "little king"; its Roman name, Regulus, was derived in the same fashion. The basilisk was considered the king of serpents (or reptiles) in the same fashion as the lion was the king of four-footed beasts.

The Basilisk in Antiquity

In the classical literature, Pliny's account of this snake-like, lethal basilisk was repeated by several other writers. In his *Theriaca,* Nicander described it with great eloquence. The warning hiss of the basilisk induced all other serpents and vermin to leave the offal they were consuming and to seek protection in their dens, leaving the basilisk to its hideous repast. When sated, it gave another sibilant hiss, but the other reptiles did not dare eat the food that had been touched by their venomous king. In Heliodorus's famous novel *Aethiopica,* the reality of such phenomena as love at first sight and the evil eye is argued from the indisputable fact that the basilisk "by its mere breath and glance will shrivel and cripple whatever comes its way." In Lucan's *Pharsalia,* the habits of the basilisk are further described:

Sibilaque effundens cunctasque tenentia pestes,
Ante venena nocens, latè sibi submovet omne
Vulgus, et in vacua regnat Basiliscus arena;

In translation:

The hissing of the basilisk all serpents terrify:
knowing of its venom, away from it they fly;
No living thing comes near it: like a hidden hand,
the loathsome beast will rule the desert sand.

A snake-like basilisk. From Edward Topsell's *Historie of Four-Footed Beasts*, published in 1607.

The basilisk had such poison within it that it could kill both humans and beasts merely by looking at them. If a valiant Moorish hunter, armed with a lance and seated on his swift Arab horse, was bold enough to attack the monster, he did not have long to triumph after having speared it, according to Lucan:

> *Quid prodest miseri, Basiliscus cuspide, mauri,*
> *Transactus? Velox currit per tela venenum,*
> *Invaditque manum;*

In translation:

> *What boots it, wretched Moor, that thou hast slain*
> *the basilisk, transfix'd him on the sandy plain?*
> *Up through the spear the subtle venom flies,*
> *The hand absorbs it, and the victor dies.*

The only way for the moor to save his life was to cut off his hand before the basilisk's venom spread to the rest of his body.

A safer and, according to Pliny, more effective way of basilisk hunting, was to put a weasel down the creature's lair, the location of which was apparent from the blackened, withered surroundings, and the thrown-up earth burned as if by a fire. The weasel and the basilisk had a natural antipathy for each other, and, by means of eating rue, the weasel could make itself impregnable to the poison of its formidable opponent. The basilisk was promptly "ferreted out" of its lair and took to flight like a weakling, but the weasel mercilessly pursued and killed it. Aelian does not agree with Pliny that the frail weasel had the power to conquer the basilisk, and instead suggested that the crowing of a cock would keep the monster at a safe distance; travelers to Africa sometimes brought roosters with them as protection against the desert basilisks. Another old legend claimed that the lion was also frightened by the cock's crowing, but the African travelers who trusted the frail and timorous birds to protect them against a pack of hungry lions were unlikely to return to tell others about their mistake.

According to the *Codex Alexandrinus,* Alexander the Great once encountered a basilisk while on his way to India with the Macedonian army. By means of a clever ruse of war, he personally put this loathsome creature permanently out of action. While conquering India, Alexander and his men were traveling through a mountain pass when, suddenly, the soldiers fell to the ground stone dead, one after the other, without being hit by enemy projectiles. When Alexander and his scouts climbed a rock to survey the surrounding area, they espied a foul basilisk standing outside its den, killing the Macedonian soldiers with its lethal stare. The situation was a precarious one, but Alexander had not studied the legend of Perseus and Medea without learning useful tricks for use against monsters such as this one; it is apparent that already at this time, people benefited from a thorough knowledge of the classics! He ordered a large shield to be polished bright as a mirror. Armed with this shield, he ran toward the basilisk's den, into which the loathsome beast had taken protection from the barrage of stones thrown by the soldiers. When the basilisk emerged from its lurking hole, the first thing it saw was its own reflection; the foul beast received such a shock that it expired there and then.

The Basilisk of the Bestiaries

In early medieval time, Europe lost contact with Africa, and the snake-like basilisk living in the Cyrenaican desert joined the ranks of other mythi-

A cock-like basilisk. From the *De Labyrynthο* of Johann Stabius.

cal monsters of old. The highly limited zoological knowledge of these dark ages was summarized in the so-called bestiaries, which were descriptive and illustrated works on the traits of various animals, often in the form of moralizing fables. In these works, the taxonomy of the larger reptiles was simple and straightforward: the dragon had wings and four legs, the basilisk had wings and two legs, and the snake (aspis) entirely lacked extremities. The basilisk further differed from the two-legged wyvern in having the head of a serpent instead of that of a cock, and the feet of an eagle. The basilisk's supernatural gifts are further expounded on in the bestiaries: the trees wither and their fruits rot when the loathsome beast passes them by, and every living thing dies in agony when struck by its death-darting glance. When the basilisk looked upward, the birds fell from the skies, like ripe plums, into the gaping jaws of the loathsome beast.

Although Pliny and Aelian had described the basilisk as a smallish snake, the early bestiary writers disagreed. They advocated a novel theory of the generation of basilisks, which made the birth of this abnormal creature as anomalous as its habits of life; this theory was first mentioned in the twelfth-century bestiary of Alexander of Neckham and is the subject of a lengthy discourse in the early-thirteenth-century bestiary of Pierre de Beauvais. An aged cock, which had lost its virility, would sometimes lay a small, abnormal egg. If this egg is laid in a dunghill and hatched by a toad, a misshapen creature, with the upper body of a rooster, bat-like wings, and the tail of a snake, will come forth. Once hatched, the young basilisk creeps down a cellar or a deep well to wait for some unsuspecting man or beast to come by, and to be overcome by its noxious vapors. These medieval writers emphasized that the basilisks occurred not only in the African desert but also in the heart of Europe: they were prevalent in Italy and France, and in the German forests. African travelers were still advised, however, to dress in leather outfits, which protected them against the basilisk's venom, and to bring weasels and cocks with them, to set on the loathsome beasts.

This medieval, cock-like version of the basilisk was also known as a *cockatrice,* a word originating as a misinterpretation of "cocodrillos" (crocodile); it was first used at the end of the fourteenth century as a translation of the Latin "basiliscus." By the High Middle Ages, the basilisk had also become firmly established as a religious symbol, signifying the devil. Some theologians, among them the learned Eusebius, speculated that the basilisk was the reptile that had tempted Eve, "as he is most venomous, and king, as it were, of the serpents." Other theologians objected that no human being could ever be tempted into accepting anything from such a deformed, pestiferous creature. The Holy Scripture has several references to "cockatrices." In Isaiah (11:8), it is said that "the weaned child shall put its hand in the cockatrice's den," while Jeremiah solemnly declares that "I will send serpents, cockatrices, among you, which will not be charmed; and they shall bite you." Many philologists are of the opinion, however, that the "cockatrice" of the Bible is a mistranslation of a Hebrew word signifying a particularly poisonous breed of snake. The basilisk is frequently found in medieval ecclesiastical art. In northern European church frescos, a popular motif is a woman in hell suckling two basilisks as a punishment for adultery; these gruesome sketches were indeed likely to remind the parishioners of the importance of a moral and decorous life. The basilisk is not uncommonly encountered among various grotesque church decorations, in France and Italy as well as in northern Europe. Carpaccio's painting of St. Tryphonius subduing the Basilisk does not portray a real basilisk, however: the creature has four legs, a lion's body, and the head of a mule. Either Carpaccio's knowledge in basilisk physiology was highly defective, or the creature in the painting was merely meant to be

A grim-looking basilisk, from John Jonstone's *Historia Naturalis*, published in 1657.

The basilisk at the portal of the church of Källs Nöbbelöv, Sweden; reproduced with permission of the Historical Museum of Lund University.

some kind of grotesque monster; it is another possibility that the name of the painting was added at a later date. A remarkable early representation of the basilisk in church art is present on a granite portal tympanum from the old church of Källs Nöbbelöv in southern Sweden, made by the master sculptor Tove in 1180. It portrays a typical medieval basilisk, with a cock's head, wings, and the winding tail of a snake, being subdued by St. Michael, who thrusts his spear into its gaping beak.

Basilisk Powder

Not the least curious part of medieval basilisk lore was the trade in basilisk powder. This substance, which was made from ground basilisk carcasses, had several uses. First, it was widely used in alchemy. "Spanish gold" could be made by treating copper with a mixture of human blood, vinegar, and basilisk powder. In the confused and fantastic imagery of English alchemy, the term *basilisk* or *cockatrice* was not infrequently used, signifying either plain evil or, through the basilisk's strange upright posture and power over life and death, eternal fame and immortality. In George Ripley's fifteenth-century *Compound of Alchymie*, a magical "Oyle" is

> *. . . made by Crafte a Stone Celestyall*
> *Of Nature so fyrye that we yt call*
> *Our Baselysk, otherwyse our Cokatryce,*
> *Our great Elixir most of pryce.*
>
> *Which as the syght of a Basylysk hys object*
> *Kylyth, so sleuth it crude Mercury,*
> *When thereon itt ys project,*
> *In twynke of an Eye most sodenly.*

Basilisk powder had several other uses. The salesmen advocated its use to exterminate vermin of all kinds, and to keep the house clean. Using superlatives reminiscent of present-day television advertisements for various cleaning agents, they pointed out that the corpse of a basilisk, hung up in the Temple of Apollo, had kept this temple free from spiders and vermin in all times, and that the carcass of another basilisk, which had been suspended in the temple of Diana, prevented any swallows from desecrating this holy place. If a house was rubbed with a pinch of basilisk powder, all swallows, spiders, and serpents were driven away. Basilisk powder was also used in medicine as an antidote for poisons, and in art: a dash of basilisk powder produced a bright red paint much favored by artists. The Spanish Moors were Europe's major dealers in this bizarre substance, supplying it to a network of salesmen and "pushers" who sold it on to alchemists, artists, medical men, and lazy housewives. There was much speculation as to how the powder was produced. According to a thirteenth-century German manuscript, the clever Moors fabricated it in underground "basilisk farms." Old cocks were kept in an underground stone cavern and induced to lay eggs by being particularly well fed. The eggs were hatched by toads and put in large copper vessels, wherein the basilisks developed. The openings of the vessels were sealed, and they were buried in a field of loose earth, the moisture of which was allowed to penetrate through a great number of small holes in the vessel. After they had been "fattened" in this way for 6 months, the copper tubs were now crowded with fully-grown basilisks. The vessels containing these loathsome beasts were then "toasted" on a big fire for several days, and the remains of the basilisks inside were ground to ashes.

The Basilisks of Rome, Basle, and Vienna

In the early 1400s, under the pontificate of Leo IV, a basilisk was found under the arch near the Temple of Lucia in Rome. Its noxious vapors afflicted the Romans with a terrible plague, but the holy pope slew the loathsome beast with his prayers. At about the same period, according to the chronicles of Laevinus Lemnius, two old cocks were discovered at Zierichzee in Zeeland, each sitting on an egg. A large mob assembled, fearing that the basilisks would be hatched and later lay waste the entire province. The cocks were beaten away from their eggs with sticks, and the pious Dutchmen later solemnly strangled these abnormal birds, in the presence of several clerics, and smashed the basilisk eggs.

In 1474, another old cock was discovered laying an egg in Basle. This provoked another outburst of the fear of basilisks: the bird was captured, jailed, and tried for this outrage against nature. In a lengthy speech, the prosecutor pointed out that the criminal rooster had threatened the whole town by endeavoring to hatch forth a loathsome basilisk. Even if a basilisk was not seen to appear, an egg of this kind, laid by a cock, was invaluable as an ingredient in various magical concoctions used to conjure up evil spirits; the laying of such an egg thus assisted the powers of evil. The cock's barrister had a difficult case to argue, since the rooster had popular opinion against it to a marked degree; its every crowing was interpreted as the confession of novel vices. The attorney stated that the bird admitted the act of laying the egg but that it was an unpremeditated and involuntary action of its body; thus, the rooster had violated no law. The clever prosecutor secured a conviction in this courtroom drama by invoking the case of the Gadarene swine: animals could be entered into by a devil and should then be destroyed. The judge solemnly convicted the rooster to be burned alive at the mountain of Kohlenberg, a place of execution reserved for particularly hardened criminals and blasphemers. A mob consisting of several thousand people had

The basilisk kills a man with its death-ray vision but is attacked by a fierce-looking little weasel. A drawing from the famous Salisbury Bestiary, reproduced by permission of the British Library.

gathered to see the cock and the basilisk egg being burned at the stake. The executioner was requested to cut the bird's abdomen open before the fire was lit. He did so, and it was apparent that it had three more eggs, in different states of development, within its oviduct. The people gave a great cry when they learned of this miracle: the pyre was lit, and when, early the following morning, the cocks of Basle greeted the sun with their crowing, only a heap of ashes remained of their abnormal, egg-laying comrade.

In the year 1202, strange things were afoot in Vienna: people fainted or were seized with violent epileptic fits while passing an old well near the center of the town. Men of observation concluded that a basilisk must be hidden in the well, poisoning it and the surrounding air with its pestiferous breath. After a long and dangerous hunt inside the dark well, the intrepid basilisk hunters were able to extract the loathsome beast from its lair; fortunately, it was already dead and could not hurt them. A statue in sandstone, depicting this zoological marvel, was erected near the well, and could be admired there for many years. A seventeenth-century writer also mentioned that a picture of the basilisk could be seen on the wall of a certain house in town. At Halle,

The frontispiece of Pierre Borel's *Historiarum et Observationum Medicorum*, published in 1676, depicting its author with a tame basilisk.

in Saxony, another ancient monument, erected to commemorate the fact that a basilisk had been seen there, was standing as late as the 1690s, and probably even longer. There were similar tales about the towns of Basle and Zwickau. In 1556, a party of workmen, who had been clearing an old well the town of Meyland, fell dead from their ladder, one by one. It was suspected that a basilisk was responsible, but although people were looking for it for several weeks, no specimen of the loathsome beast was found, dead or alive. In a French town, exactly the same thing happened a few decades later, which resulted in a veritable mass hysteria. The basilisk hunt was again unsuccessful.

The Basilisk of Warsaw

The critically minded reader might rightly object that all these observations of basilisks were of a particularly diffuse character: often, no one actually saw the loathsome beast that was supposedly lurking down in the well or vault. But there exists, indeed, one detailed and apparently convincing sighting of a basilisk; the hideous beast observed in Warsaw, in the year 1587. The 5-year-old daughter of a knifesmith named Machaeropaeus had disappeared in a mysterious way, together with another little girl. The wife of Machaeropaeus went looking for them, along with the nursemaid. When the nursemaid looked down into the underground cellar of a house that had fallen into ruins 30 years earlier, she observed the children lying motionless down there, without responding to the shouting of the two women. When the maid was too hoarse to shout anymore, she courageously went down the stairs to find out what had happened to the children. Before the eyes of her mistress, she sank to the floor beside them, and did not move. The wife of Machaeropaeus wisely did not follow her into the cellar, but ran back to spread the word about this strange and mysterious business. The rumor spread like wildfire throughout Warsaw. Many people thought the air felt unusually thick to breathe and suspected that a basilisk was hiding in the cellar. Confronted with this deadly threat to the city of Warsaw, the senate was called into an emergency meeting. An old man named Benedictus, a former chief physician to the king, was consulted, since he was known to possess much knowledge about various arcane subjects. The bodies were pulled out of the cellar with long poles that had iron hooks at the end, and Benedictus examined them closely. They presented a horrid appearance, being swollen like drums and with much-discolored skin; the eyes "protruded from the sockets like halves of hen's eggs." Benedictus, who had seen many things during his 50 years as a physician, at once pronounced the state of the corpses an infallible sign that they had been poisoned by a basilisk. When asked by the desperate senators how such a formidable beast could be destroyed, the knowledgeable old physician recommended that a man descend into the cellar to seize the basilisk with a rake and bring it out into the light. To protect his own life, this man had to wear a dress of leather, furnished with a covering of mirrors, facing in all directions.

Benedictus did not, however, volunteer to try out this plan himself. He did not feel quite prepared to do so, he said, owing to age and infirmity. The senate called on the burghers, military, and police but found no man of sufficient courage to seek out and destroy the basilisk within its lair. A Silesian convict named Johann Faurer, who had been sentenced to death for robbery, was at length persuaded to make the attempt, on the grounds that he be given a complete pardon if he survived his encounter with the loathsome beast. Faurer was dressed in creaking black leather covered with a mass of tinkling mirrors, and his eyes were protected with large eyeglasses. Armed with a sturdy rake in his right hand and a blazing torch in his left, he must have presented a singular aspect when venturing forth into the cellar. He was cheered on by at least two thousand people who had gathered to see the basilisk being beaten to death. After searching the cellar for more than an hour, Faurer finally saw the basilisk, lurking in a niche of the wall. Old Benedictus shouted instructions to him: he was to seize it with the rake and carry it out into the broad daylight. The brave Johann Faurer accomplished this task, and the populace ran away like rabbits when he appeared in his strange outfit, gripping the neck of the writhing basilisk with the rake. Dr. Benedictus was the only one who dared examine the strange animal further, since he believed that the sun's rays rendered its poison less effective. He declared that it really was a basilisk: it had the head of a cock, the eyes of a toad, a crest like a crown, a warty and scaly skin "covered all over with the hue of venomous animals," and a curved tail, bent over behind its body. The strange and inexplicable tale of the basilisk of Warsaw ends here: none of the writers chronicling this strange occurrence detailed the ultimate fate of the deformed animal caught in the cellar. It would seem unlikely, however, that it was invited to the city hall for a meal of cakes and ale; probably, the versatile old Doctor Benedictus probably knew of some infallible way to dispose of these monsters.

The basilisk and the mirror: a drawing from the 1661 edition of Joachim Camerarius's *Symbolorum et Emblematum*.

The Basilisk of Copenhagen

On Easter Sunday, 1649, the German doctor Ludwig Kepler was enjoying his breakfast, together with his family. He was a son of the famous astronomer Johannes Kepler, and a practitioner of repute. When he thrust his spoon into a boiled Easter egg, he was aghast to feel a small snake-like creature inside it! Ludwig Kepler's account of this strange happening neglects to mention if the basilisk had been hard boiled or soft boiled. He does point out, however, that this unexpected encounter with the King of Serpents at his own breakfast table did not diminish his appetite for eggs; indeed, he had eggs served at regular intervals to be examined for the presence of young basilisks within. After 8 years of futile study, Kepler consulted the famous Danish anatomist Thomas Bartholin, who was well known to be an authority on basilisks. In April 1651, Bartholin's knowledge of basilisk physiology had been put to practical use in a crisis affecting the fate of the Danish monarchy. When one of the royal cupbearers was collecting eggs from the hen coops situated within the castle of Copenhagen, he was greatly frightened to see, right before his eyes, an old cock laying an egg! The startled ganymede ran back to the royal kitchen, where his fantastic tale soon spread like wildfire among cooks, servants, and courtiers. King Frederick III of Denmark ordered that the basilisk egg be collected and closely watched for several days. The egg was smaller than an ordinary hen's egg but larger than a pigeon's egg. No basilisk appeared, however, and the egg was later put in the royal cabinet of curiosities, where it was to reside for many years. The royal chefs were much frightened that the cock had laid more eggs than one, and they broke each egg with great caution while preparing the king's breakfast. None of them had a good recipe for a basilisk omelet, and they feared that the king of all Danes would not appreciate being served the king of reptiles on his breakfast tray.

The egg-laying cock was also captured by the Danish courtiers. King Frederick ordered that it should be dissected by Thomas Bartholin in his own presence; to their surprise, they found that the cock had normal male genitals. Thomas Bartholin doubted that it could really have laid an egg by the natural route, but the kitchen servant stubbornly maintained that this cock, and no other bird, had laid the egg in question. While discussing the case in a scholarly treatise, Bartholin mentioned the possibilities that the abnormal egg was either developed from the old cock's rotting semen or that it had formed in the intestines. Ludwig Kepler suspected, however, that Bartholin doubted the veracity of the individuals involved in the scandal but did not want to disappoint his royal master, whose museum had been graced with the presence of the King of Reptiles, albeit in an undeveloped state. Frederick III also had the skeleton of the old cock prepared and mounted in his museum, where it was to reside for almost 200 years.

For many years, Thomas Bartholin pondered the riddle of the basilisk. In 1673, he encountered some other alleged cock's eggs. He was, by this time, quite unconvinced of the existence of basilisks. He incubated both normal hen's eggs and an alleged cock's egg by placing them in an alembic of heated sand. This experiment, which some of his contemporaries must have considered very hazardous indeed, did not have a fatal outcome; no basilisk leapt out of the alembic to poison the courageous Danish anatomist, but the cock's egg proved to be sterile and filled only with a whitish, gelatinous jelly. Bartholin declared that these alleged cock's eggs had, in all probability, been laid by elderly hens, which were constitutionally unable to produce normal eggs. He had himself observed the laying of these small, sterile eggs from elderly birds of that kind. What puzzled him was that there was a steady

A drawing of a basilisk, published by Georg Wedel in the *Ephemerides* of 1672.

inflow of sightings of young basilisks inside eggs. He had himself seen one in Denmark. His friend, the famous anatomist Nicolaus Steno, told him that he had recently witnessed a remarkable scandal during a stay in Florence. An egg broken to be cooked for the grand duchess of Tuscany was found to contain a living "small serpent," which darted out with great alacrity as soon it was freed from its confinement. The cook succumbed to a nervous attack, as did the kitchen maids; the court was buzzing with rumors. The duchess ordered that a body of anatomists and zoologists scrutinize the young basilisk, to determine how it had been engendered. The Danish visitor, who already held an acclaimed position in anatomical circles, was also consulted. The board of experts pronounced it likely that the basilisk had, indeed, been produced from the egg of a cock and hatched by a toad, but Thomas Bartholin doubted this explanation. He rather suspected that an amorous Italian serpent had coupled with a hen, and that the little snake within the egg was the result of this outrage against nature.

A couple of years after the basilisk of Copenhagen had joined the royal museum, Frederick III again consulted Thomas Bartholin. The king had purchased a dried basilisk from a dealer in natural curiosities, planning to exhibit it together with the cock and the egg, and thereby demonstrate all stages in the development of the loathsome beast. After closely examining the strange exhibit, Bartholin had to tell the proud royal collector that he had been duped: the "basilisk" was a fake, constructed from the body of a ray. Nevertheless, this counterfeit basilisk was exhibited in the royal museum for many years, although in the catalogue, Bartholin's brother-in-law Holger Jacobsen confessed that it was a manufactured specimen.

Cock's Eggs and Faked Basilisks

The Italian Count Ulysses Androvandi was one of the foremost polymaths of the Renaissance: he was called the Pliny of his time, and planned a huge series of folio volumes to illustrate his immense collections from the animal, vegetable, and mineral kingdoms. At Count Aldrovandi's death in 1605, only four volumes—these dealing with the birds and insects—had been published. His students worked for decades to compile eight more volumes from his immense collection of manuscripts and drafts. Like many of his contemporaries, Ulysses Aldrovandi was fascinated by the riddle of the basilisk. He described all aspects of this loathsome beast in a long manuscript that his posthumous editors incorporated into his *Historiae Serpentum et Draconum Libri Duo,* which was published as late as 1640. If a review written by a present-day frontline scientist would have been discovered 35 years after his death, it would have been difficult to get this obsolete manuscript published, but seventeenth-century basilisk research moved at a stately pace: Aldrovandi's chapter about the basilisk was hailed by many contemporary writers as the best of its kind, and it still commands respect. A man of immense learning, Aldrovandi had compiled the opinions of many obscure Latin authorities: natural historians, poets, and writers of anecdotes. He also presented his own views on the riddle of the basilisk's generation.

Aldrovandi was himself the owner of a basilisk—or at least a dried specimen once seen by the learned Mercuralis in the "treasure house," or private museum, of the Emperor Maximilian of Austria. Mercuralis accepted it as a true basilisk, but Aldrovandi saw that it had been manufactured from a dried ray. Ulysses Aldrovandi had spent much time observing the development of the egg and chicken; to this day, he is considered one of the pioneers of embryology. Although he knew that birds resembling cocks could, on rare occasions, lay eggs, he believed these eggs to be small and sterile and without a yolk; they were reminiscent of the eggs laid by old, decrepit hens. This made him doubt that the basilisk was engendered from a cock's egg. During Aldrovandi's long and varied life, he had not once seen a basilisk, living or dead, nor had any of the hundreds of scholars within his acquaintance. After this piece of deductive reasoning, rare enough in early seventeenth century basilisk studies, one would have expected that Aldrovandi would have wholly denied the existence of this loathsome beast. But he was imaginative by nature as well as reverent toward the many earlier writers who had described the basilisk in detail. Furthermore, his friend, the anatomist and embryologist Fabricius ab Aquapendente, had once given him a "young basilisk," which this learned anatomist had found in an egg that he was eating. Another, even stranger egg in Aldrovandi's collection had the imprint of a small, winding snake on its shell! Ulysses Aldrovandi was not sure that it had "the mark of the basilisk," however. He also considered the possibility that the hen had been frightened by a snake and that this experience had, by the strange workings of the maternal imagination, been imprinted on the shell of the egg.

Throughout the 1600s, eggs containing snake-like "young basilisks" turned up at regular intervals; these macabre findings were discussed in the leading scholarly journals of the time and did much to prevent the lore of the basilisk from being forgotten. In 1673, a live "little snake" was discovered in an egg broken by one of the ladies in waiting to the Duchess Dowager of Parma; once again, the King of Serpents was haunting royalty! The loathsome beast was described by Dr. Jerome Santasofia, Professor at the Parma medical school, and Dr. Jacques Grandi, anatomist of Venice. It was as long as a forefinger, with a flattened head, and as thick as a cherry stalk. Six years later, Dr. André Cleyerus described an egg that had, like the one described by Aldrovandi, "the mark of the basilisk": the relief of a winding little snake was encrusted into the shell.

In early March 1672, the Jew Isaac Lefkowitz, living in Thurn, Poland, was cooking a meal of egg soup for his large family. When broken, one of the eggs turned out to have a black, wriggling, snake-like creature inside it, to the great astonishment of Isaac Lefkowitz. He cried out in terror and threw the egg away from him. The odor from the egg was overwhelming, and his wife and young children fell to the floor, unconscious, when struck by these noxious vapors. Lefkowitz's eldest son was nearly overcome by the frightful smell, but he managed, with his last remaining strength, to kick this stink bomb of an egg out through the door. Out on the street, it was trodden underfoot by the playing children, who did not have enough sense to fear the basilisk. The distraught Lefkowitz called the anatomist Dr. Simon Schultz, who examined the flattened remains of the Polish basilisk. In his case report, Dr. Schultz cautiously discussed the possible explanations of this *lusus naturae:* either had it been a cock's egg, and thus a genuine basilisk, or else the hen laying the egg must have eaten a snake's egg, whose molecules had, in some strange and disastrous manner, mixed with those of her own chicken.

In 1671, the Jena physician Friedrich Madewisius published his doctoral thesis on the basilisk. In the manner of the time, he quoted the opinions of a multitude of earlier writers on this subject. The conclusion is that, although no one had ever seen a basilisk and lived to describe it, it could not be doubted that this monstrous creature really existed. The question of its generation was more difficult: Madewisius quoted various theories, among which he supported the version that it was spontaneously generated from the putrid, infected effluvia inside the egg laid by a decrepit old cock. Another German thesis on basilisks was published in 1697, after a certain Dr. Johannes Haenfler had had occasion to examine the egg of a peacock, laid in the yard of Judge Prentzlo, in sight of his servants' quarters. It had been widely exhibited as a prodigy.

When Thomas Bartholin exposed the faked basilisk of Copenhagen, he presented one of his arguments that he had seen several similar, artificial specimens during his travels on the continent. Many seventeenth-century naturalists had a dried "basilisk" in their collections, although some, such as Ulysses Aldrovandi, were aware that it was a fake. Aldrovandi also had a dried "dragon," which he figured in the *Historiae Serpentum et Draconum,* with the added comment that "showmen fashion diverse-shaped figures from dried rays and foist these on the ignorant either for dragons or for basilisks." Not a few royal museums were among the buyers: the wealthy and ignorant princes were pleased to pay huge sums to purchase a specimen of the King of Serpents for their collections. Although King Frederick's faked basilisk is no longer in existence, several other specimens are still preserved, many of them in Italy. There are basilisks in both Milano and Venice, and the Museo di Storia Naturale in Verona has not less than three. One of these was purchased in the early seventeenth century by Signor Francesco Calzolari, a wealthy Veronese apothecary who was a correspondent of Ulysses Aldrovandi. Already in the 1622 catalogue of this museum, the basilisk had been exposed by two skillful naturalists who declared that "this monstrous animal . . . is not a Basilisk, nor a Dragon, but made from a fish called the Ray, by artificial means."

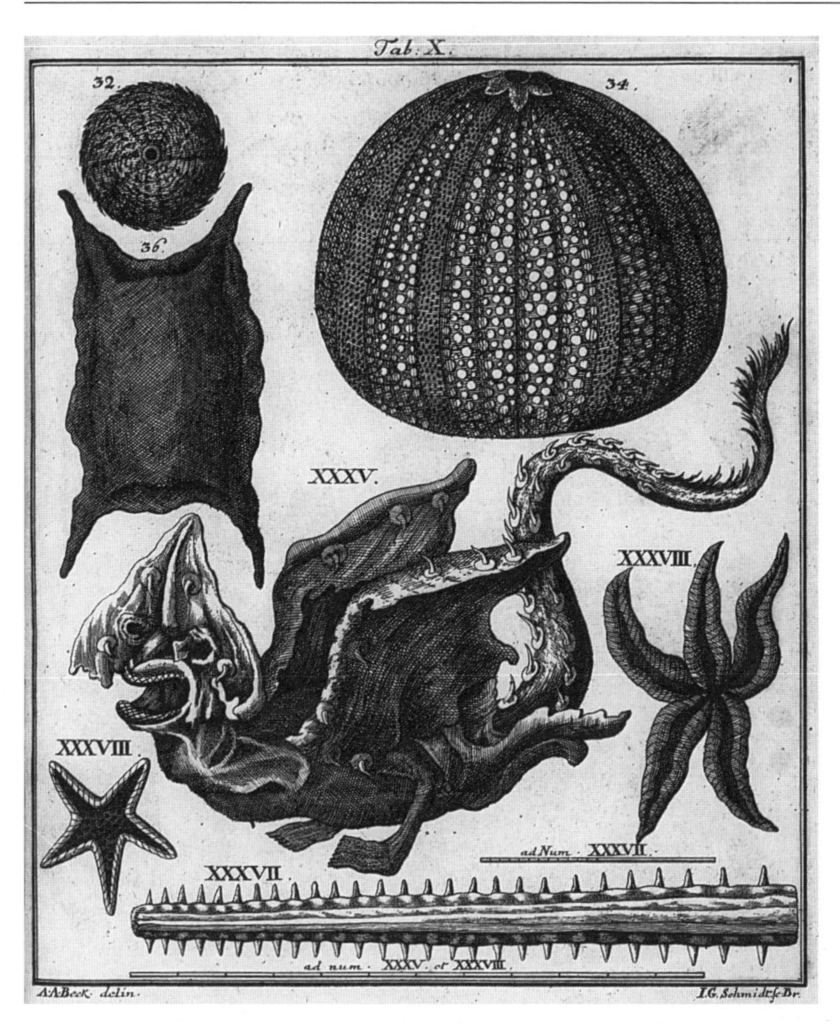

A faked basilisk, which is one of the specimens in the cabinet of curiosities of the brothers Besler, as described by J. H. Lochner von Hummelstein in 1716.

In his 1716 catalogue of the cabinet of curiosities belonging to the brothers Besler, J. H. Lochner von Hummelstein described another dried basilisk: a winged, four-legged monster, which is also likely to have been fabricated from a ray. In his *Traité Générale des Pesches,* the French zoologist H. L. Duhamel du Monceau described another quite artistic basilisk that had been constructed from a dried lizard and a ray. At least one faked basilisk was on show in London. According to a rare handbill kept at the British Library, entitled "A Brief Description of the Basilisk, or Cockatrice," and containing a crude engraving of a merry basilisk killing three men with its "death ray," it was the property of "James Salgado, a Spaniard and converted Priest." The basilisk had been purchased from a reputable Dutch physician. Mr. Salgado may have been the author of the Latin doggerel:

Quos vivens vidi, nunc mortus.
Uni de vitam, dum me gens numerosa videt.

which was translated as

All men I kill'd that I did see
But now I am dead one lives by mee.

Poor Mr. Salgado did not live well off his basilisk, however; according to the handbill, he brought it before his "Honourable Benefactors" in London to implore them to "help a distressed stranger who is reduced to very great straights."

Shakespeare's Basilisks and Cockatrices

According to Edward Topsell's *History of Four-Footed Beasts,* Britain was once considered a particular haunt of basilisks. An old tradition said that a brave (and inexhaustible) man once wandered through every road of the entire British Isles, covered with a multitude of mirrors; when these noxious animals saw themselves or each other, they all expired. This bold basilisk exterminator seems to have been remarkably successful: except for the pathetic Mr. Salgado's dried specimen, there are few reports of basilisks invading Britain. It is notable, however, that Samuel Pepys's *Diary* mentions that some serpents in Lancashire, which "grow to a great bigness," are able to kill larks by using their death-ray vision, making the birds fall down from the sky into the jaws of the waiting snake. This tale may well have been a remnant of an older basilisk tradition. Otherwise, the British basilisks and cockatrices were encountered only in contemporary literature and poetry, where they appear in many guises. The word *basilisk* is mentioned not fewer than 112 times in the English poetry database, from medieval times until the year 1900. William Shakespeare's plays abound with basilisks, which are regularly used to symbolize the evil eye. In the second part of *Henry VI,* the king accuses Suffolk for Gloucester's murder in an impressive speech:

Upon thy eyeballs murderous tyranny
Sits in grim majesty to fright the world.
Look not upon me, for thine eyes are wounding:
Yet do not go away; come, basilisk,
And kill the innocent gazer with thy sight;
For in the shade of death I shall find joy,
In life but double death, now Gloucester's gone.

In *Richard III,* the hunchbacked, evil future king courts the Princess Anne at her late husband's funeral. His insinuating compliment, "Thine eyes, sweet lady, hath infected mine," has little success with the fair one, however, and she answers him, "Would they were *basilisks,* to strike thee dead!"

William Shakespeare was not the earliest writer to use the basilisk as a metaphor. Already in Geoffrey Chaucer's *Personne's Tale,* appears the sentence: "That sleeth right as the Basilicok / sleeth folk by the venym of his sighte." And in Caxton's *Jason,* it is said that "Certes madame your eyen / basilique have hurte me unto the death." In Pope's *Messiah,* there are instead shades of the Biblical basilisk: "The smiling infant in his hand shall take / The crested basilisk and speckled snake."

Josiah Sylvester's translation of Guillaume de Salluste du Bartas's *La Semaine* revives the traditional image of the basilisk:

What shield of Aiax could avoid their death
By th' Basilisk, whose pestilentiall breath
Doth pearce firm Marble, and whose banefull ey
Wounds with a glance, so that the soundest dy.

The same poet deserves to be honored as an early enemy of tobacco smoking. In 1614, he published a witty poem appositely named *Tobacco Battered, or the Pipes Shattered About Their Eares that Idely Idolize So Base . . . a Weed,* containing the lines:

If then Tobaccoing *bee good: How is't,*
That the levdest, loosest, basest, foolishest,
The most unthrifty, most intemperate,
Most vitious, most debaucht, most desperate,
Pursue it most: The Wisest and the Best
Abhor it, shun it, flee it, as the Pest,
Or pearcing Poison of a Dragons whisk
Or deadly Ey-shot of a Basilisk;

Today, tobacco has definitely taken over the role of dragons and basilisks as a serious threat to public health.

In Beaumont & Fletcher's *Philaster,* the basilisk is used as an invective against women:

> *Let me love lighting, let me be embraced*
> *And kissed by scorpions, or adore the eyes*
> *Of basilisks rather than trust the tongues*
> *Of hellbred women.*

Even stranger: the word *cockatrice* was commonly used, throughout the seventeenth century, to mock a loose woman, particularly if she was the kept mistress of some military officer. In Ben Jonson's *Cynthia's Revels,* it is said that "No courtier but has his mistress; / No captain but has his cockatrice." And in John Taylor's works, a woman is described with the following words: "And among souldiers this sweet piece of vice / Is counted for a captain's cockatrice."

John Dryden alluded to the later, garbled version of the basilisk legend, that the foul beast could kill only by looking at some object, and that it died when looked upon: "Mischiefs are like the cockatrice's eye; / If they see first, they kill; if seen, they die." Sir Francis Bacon used the same metaphor in *Henry VII:* "This was the end of this little cockatrice of a King that was able to destroy those that did not espie him first." In his *Robespierre,* Samuel Taylor Coleridge likened this French tyrant to "The crowned cockatrice whose foul venom / Infects all Europe."

The End of the Basilisk

In the early 1600s, knowledge in natural history was organized in the same way as in the Middle Ages. In the massive zoological treatises of Conradus Gesner, Ulysses Aldrovandi, and Edward Topsell, much material was fetched from Pliny and the medieval bestiaries. Proper knowledge of domestic animals was intermixed with far-fetched tales of strange monsters; pigs, hens, and horses appeared side by side with mantichoras, mermaids, and unicorns. The majority of naturalists were still in favor of the basilisk's existence. One of the few dissidents was the French naturalist Jean Bodin, who asked himself, "If the basilisk kills by being seen, then who has ever seen it"; his learned adversaries solemnly reminded him that it was, since Pliny's time, common knowledge that the basilisk killed while seeing its foe, not merely by being seen.

In the sturdy tomes of John Jonstone and Edward Topsell, basilisk lore is thoroughly reviewed. Jonstone made the telling remark that the descriptions of this loathsome beast by different authors through the ages were so dissimilar that they might reflect several different species of animals. Sir Thomas Brown discussed the riddle of the basilisk at length in his *Pseudodoxia Epidemica.* He was bold enough to doubt the old fable that the basilisk "proceedeth from a Cock's egg hatched under a Toad or Serpent," considering this "a conceit as monstrous as the brood it self." Although he did not doubt that many strange and venomous reptiles could be engendered in this way, he doubted that this abnormal form of generation would always produce a basilisk. Nor was he disposed to accept the basilisk's "death ray" through its vision, since, like contemporary physicists, he correctly believed "that sight is made by Reception, and not by Extramission; by receiving the raies of the object into the eye, and not by sending any out." In spite of these objections, he was impressed by the testimony of many earlier authorities, including the Holy Scripture, and lamely had to admit "that such an animal there is . . . we cannot safely deny."

No living basilisks were encountered during the entire seventeenth century: the horrible monster of Warsaw seems to have been the ultimate one. This dearth of actual observations of basilisk did not stop imaginative naturalists from discussing the loathsome beast; it lived its own secluded half-life in obscure scholarly works for yet another decade and a half. The Jesuit father Athanasius Kircher was one of the basilisk's staunchest defenders. In his *Mundus Subterraneus,* published in 1664, he declared that it would be considered very bold indeed for any individual to deny that basilisks existed, and that they were engendered from a cock's egg. He had heard from his Jesuit brethren at Perugia that, in February 1661, a cock had laid an egg from which a terrible basilisk had burst forth: it had wings, sharp teeth, and a long, curved horn on its forehead. It is left tantalizingly unclear how the people disposed of this formidable monster. Athanasius Kircher had himself never seen a basilisk, dead or alive. The nearest thing was a strange cock living in the garden of the grand duke of Tuscany. It had a snake-like tail, long legs, and a large comb like a crest. Some people believed it to be a tame basilisk, and they ran away when the grand duke's servants put a leash on the bird and took it for a walk around the park.

The German Eberhard Happel incorporated a long section on basilisks in his *Relationes Curiosae,* published in 1687. He took the loathsome beast quite seriously and discussed the evidence for and against its existence with Teutonic precision. Another German naturalist, Abraham Theodor Krafft, again discussed the basilisk at length in a popular early–eighteenth century manual about how to effect "The complete Extermination of Animals Dangerous to Men and Cattle." As judged from his advice of various cruel methods to eradicate household pests, Dr. Krafft was no lover of animals, but he was full of respect for the basilisk, whose supernatural talents made it the nightmare of an exterminator of vermin. As late as 1736, a complete zoological treatise on the basilisk, written several years before by the prolific naturalist Georg Caspar Kirchmaier, was reprinted in both Latin and German and was later translated into English. The early modern zoologists, such as Linnaeus, Buffon, and Blumenbach, were more difficult to convince about the basilisk's existence. Like many of the fabulous animals from the bestiaries, such as the dragon, the mantichora, and the griffin, the basilisk was an early casualty of the Age of Reason. In the mid-1700s, no basilisk had been sighted for more than 150 years, and even traditionalist theologians had to accept that even if there had once been such animals, they were no longer walking the earth.

Long after the King of Serpents had been dethroned from its prominent position in zoological taxonomy, it still reigned supreme in folklore. Ethnologists have traced basilisk lore to all parts of Europe: it was prevalent in Poland, Germany, and Scandinavia and could also be encountered in Ireland, Spain, France, and Italy, and even in Arabia and among the Chinese. In particular, the idea of the development of a basilisk, or "dangerous worm" from a cock's egg seems to have been extremely widespread. It was considered a bad omen if a hen crowed like a cock: the farmer must kill this unnatural bird and not suffer it to produce another egg. This old belief is the subject of an English proverb, "A whistling girl and a crowing hen / Are neither fit for gods nor men," as well as a German one:

Wenn die Henne kräht vor dem Hahn
Und das Weib redet vor der Mann
So soll man die Henne braten
Und das Weib mit Prügeln berathen;

Neither leaves any question as to which is the superior sex!

There were also regular outbreaks of basilisk hysteria, after a cock had been observed to lay an egg. This had been judged to be impossible by the eighteenth-century naturalists, but neither the cocks nor the common people heeded this anathema. In 1730, a Swiss cock was executed for such a crime against nature, and as late as 1838, an Irish cock faced trial after having been observed to lay several eggs. After reliable farmers had testified as to its unnatural behavior, the bird was sentenced to burn at the stake, along with one of its eggs. As the pyre was burning fiercely, the superstitious people believed that the egg burst asunder and that a dreadful creature like a serpent came forth. The need to bring these wicked and scandalous birds to justice was further emphasized by the discovery, at regular intervals, of eggs containing "young basilisks": living, snake-like creatures. These bizarre eggs kept turning up long after the rest of the basilisk lore had been laid to rest. When, in 1810, an egg laid by a duck belonging to Mr. Clemenshaw, of Winmoor, near Leeds, was broken up to be cooked for dinner, a "young snake," stated to be 10 inches in length, fell out. According to the *Canadian Courant,* when a lady broke an egg to cook it in an omelet in November 1827, a small snake, 2 inches in length, burst out and crawled about the frying pan with much avidity.

A strange variation of the basilisk legend, occurring in Scandinavia and Germany well into the nineteenth century, was that basilisks could be engendered in old mead barrels that had been kept for a long time without being cleaned. The fermenting dregs of these huge barrels condensed into a firm lump of matter called a "slårk"; this noxious corpuscle could, if allowed to ferment for a long time, develop into a basilisk. In a tale from Frederikshavn, Denmark, a very old barrel that had been kept for 100 years was taken out of a cellar. Optimistically, to say the least, the Danish brewer sent a serving girl to open the tap to find out if it still contained any beer worth drinking. She heard something move inside the barrel, however, and alerted her fellow workers, who feared that a basilisk had been formed inside the barrel. The master brewer ordered that the barrel be taken outside town to be burned on a pyre. As it burned up, they saw the basilisk leap high in the air, but failed to save itself from the flames.

The Riddle Solved

In antiquity, the North African desert made a profound impression on the early Greek and Roman travelers: they had never seen such a withered, barren landscape, and there was much speculation as to the cause behind the formation of these huge wastelands. One hypothesis was that the basilisks and other venomous reptiles inhabiting the Cyrenaican deserts had destroyed them by their noxious fumes. The "basilisk" described by Pliny much resembles a still extant desert serpent named *Lytorhynchus diadema.* As in Pliny's description, it is about 6 inches in length, and yellowish brown in color, with a white "diadem" on its head. This harmless reptile was blamed for the vegetation becoming blackened and shriveled in the sweltering heat, and the rocks bursting by the night frost in winter. Pliny's description of the basilisk serpent might also have been inspired by early accounts of the natural history of cobras. These snakes can raise their heads and bodies from the ground when teased or enraged, and they sometimes crawl with their heads upraised like the basilisk; the "spectacle" sign on their neck shields may have inspired the tale of the basilisk's royal crest. The great skill of the mongoose in fighting the cobra and other venomous serpents may have inspired Pliny's account of the enmity between the weasel and the basilisk. This tale also illustrated his own theories about the equilibrium of nature: every animal, even the most terrifying monster, must have a natural enemy capable of destroying it.

It is not unlikely that the lore of the snake-like basilisk of antiquity can be linked to Egyptian mythology; it may thus have been established many centuries before Pliny's account. An important clue is given by the *Hieroglyphica* of Horapollon, which likens the Greek *basiliscus* to the Egyptian ouraion snake. He further writes that if the ouraion snake breathes on any living being, it dies, and the snake does not have to bite it. The ouraion snake was a royal attribute among the Egyptians. Indeed, the snake emblem adorning the brows of the Egyptian pharaohs is an ouraion snake, which protects the pharaoh with its fiery breath.

The legend of the basilisk hatched from a cock's egg can be traced back to the tenth century but no further. During antiquity, it was even thought that the crowing of the cock drew away the African basilisks. A possible source of both these versions was Herodot's fanciful tale of the winged serpents of Araby, which every year invaded Egypt in great numbers. They were defeated by the ibis birds, which attacked the serpents and ate them whole. The Egyptian peasants were grateful to their native birds, but the birds' heroic act had brought about lurid changes in their reproductive physiology: through eating the venomous serpents, the ibises themselves became completely saturated with poison and laid eggs containing young serpents instead of chicks. This may well have been the origin of the medieval tale of the basilisk coming from the cock's egg. The observations of Ulysses Aldrovandi and Thomas Bartholin did much to discredit this old belief, but it remained until the early eighteenth century. In 1706, the French anatomist Lapeyronie was instructed by the French Academy of Sciences to examine an alleged egg-laying rooster. The eggs in question were plainly laid by an elderly hen. Lapeyronie concluded that cocks were constitutionally incapable of laying eggs, and that the small, abnormal "cock's eggs" shown to him were instead produced by old, sickly hens. The majority of later writers have agreed with him. When the French zoologist C. Davaine wrote a lengthy review about anomalous eggs in 1860, for example, he considered all the old reports of basilisk embryos inside eggs to be wholly imaginary, and due to an overheated imagination.

In 1780, the great British surgeon and anatomist John Hunter described an "extraordinary Pheasant": it was a pheasant hen that had, in old age, acquired the plumage of a cock. Several other eighteenth-century anatomists made similar observations, Buffon, Blumenbach, and Geoffroy St. Hilaire among them. The phenomenon itself had already been described by Aristotle, and Livy wrote that among the portents observed just before Hannibal's invasion of Italy in 217 B.C. were "the changing of a hen into a cock and a cock into a hen." Modern ornithologists agree that spontaneous intersexuality among birds really does exist but that it is possible only for hens to acquire male sexual characteristics: several twentieth-century cases have been described, in chickens as well as in wild fowl. The physiological background for this is that birds, like human beings, possess latent male and female secondary characteristics, the expression of which is controlled by the production of male or female sex hormone. Just as a woman with an adrenal or ovarian disorder that leads to the production of an excess of male sex hormone may undergo virilization, a hen bird can be the victim of a similar disease in the ovary. The balance of female and male sex hormone is disturbed, causing the hen to develop a rooster's coat of feathers. This phenomenon is sometimes observed in a decrepit old hen, whose ovary atrophies in old age; the production of male sex hormone from the rudimentary right gonad takes over and the bird undergoes partial or total virilization, even usurping the male prerogative of crowing. The oviduct and ovary of these aged "cocks" are quite atrophied, and these animals are incapable of laying eggs. It has also been observed, however, that hens in the prime of life may change their sex owing to some disease that reduces the production of female sex hormone from the ovary. This might be caused by a tumor or cyst in the ovary, or by some kind of granulomatous inflammation. These remarkable "roosters" may have fully developed secondary male characteristics but are still capable of laying eggs. It has even been recorded that one, unique hen that was previously the mother of several chicks became fully masculinized in behavior, fertilizing the eggs of a normal hen and becoming the father of chicks also!

Another remarkable observation was made by the Canadian zoologist L. J. Cole in 1927. A brown leghorn "rooster" had laid several eggs, which were of normal size and structure. The animal's comb and wattles were rather like those of a hen, however, although the plumage was typically male in character. The sex change is likely to have occurred swiftly: some feathers from its former coat were remaining, and they were those of a hen. The bird was mated with a white leghorn rooster, which must have felt rather confounded, not to say dismayed, when viewing his prospective bride, although this did not prevent nature taking its course and a number of normal chicks being raised. Remarkably, the egg-laying rooster regained its female coat of feathers half a year later, and it returned to its uneventful life at the Wisconsin egg central, without any apparent ill effects of its brief masquerade in the plumage of a male. The bird was never autopsied, and one can only speculate that it must have had some self-limiting inflammatory or granulomatous disease of the ovary. The famous egg-laying rooster of Basle, which suffered a hotter death than in the stew as a punishment for its crimes against nature, is likely to have suffered from a similar disorder.

The fact that these legendary oviparous roosters really did exist explains one of the most puzzling parts of the riddle of the basilisk. Another mysterious feature of the generation of that loathsome beast was the occurrence of eggs containing "young basilisks." These abnormal eggs were described at regular intervals during the seventeenth and eighteenth centuries: in all, more than fifteen observations exist, many of which were published in the scholarly journals of the time. In the nineteenth century, these bizarre findings were doubted by ornithologists and natural historians and have even been used by modern writers to exemplify the ignorance and credulity of the old seventeenth-century writers on natural history. Some zoologists have suggested that the old "basilisk eggs" were in fact eggs of the grass snake, but it is an egregious insult to competent natural historians such as Ulysses Aldrovandi and Thomas Bartholin to suggest that they could not tell the egg of a grass snake from that of a hen. Another, slightly more reasonable hypothesis is that the "basilisk embryos" were in fact the so-called chalaza parts of the egg: string-like structures that connect the yolk of the egg to the poles of the shell. In an abnormal egg, which lacks the normal yolk, the chalaza structures may be fused and thereby resemble a writhing little snake. The zoologists who have followed Lapeyronie in advocating this hypothesis would have difficulties in explaining, however, why the "young basilisks" were often found in eggs with normal yolks, in which the chalaza structures are divided. Furthermore, seventeenth-century observers often reported that the basilisks were alive and that they sometimes moved about with great vigor!

The answer to this riddle may seem even more bizarre than the two theories reviewed earlier. Unlike the other theories, the latter explanation is supported by several reliable twentieth-century observations of the phenomenon in question. Hens, like humans, occasionally harbor intestinal worms within them. The largest of these are roundworms such as *Ascaris lineata.* These ascarid worms, which may become 1, or even 2, inches long, live their natural lives within the hen's gut, but they may also crawl out into the so-called cloacum, where the gut and the oviduct join. Usually, they later follow the hen's droppings out into the open. Some worms, however, inspired either by a desire to explore their surroundings or a poor sense of direction, might actually take a wrong turn, and instead travel up the oviduct, to be included in the fowl's egg. If a worm reaches the isthmus part of the oviduct, it is incorporated into the white of the egg, and will constitute a very unpleasant surprise for the individual trying to eat it. If it has not reached as far as that, it is instead incorporated into the eggshell; this explains the eggs with "the mark of the basilisk" observed by Aldovandri and Cleyer. A close study of the old and new literature thus vindicates the veracity of the old naturalists: the "basilisk embryos" observed inside hen eggs were no mere figments of the imagination but rather large roundworms incorporated into the white of the egg.

Although modern egg eaters may find such an occurrence distasteful, several instances of this unsavory phenomenon have been described by twentieth-century ornithologists. The "basilisk eggs" still occur, although, like other impossible or "damned" observations that are supposed to be contrary to the established order of nature, such as spontaneous human combustion and rains of live frogs, they are given little attention by naturalists. According to a story published in many American newspapers, Ms. Ursula Beckey of Long Island cracked some eggs to make an omelet in October 1979. Out of the third egg crawled a 3-inch "snake"! Four months later, she filed 3.6-million-dollar damages against the shop that had sold her the eggs, claiming that the shock might well have killed her, and that she now felt nausea just looking at an egg. Her attorney unfortunately got hold of the wrong end of the stick, to put it mildly: his expert comment on this strange case was that "I have been told that a snake of this size in an egg must have come from a mother snake at least six feet long"!

Many historians of zoology have marveled that the basilisk—this protean monster, depicted in many different shapes and forms—could remain within the texture of seventeenth-century zoological taxonomy, and that it even survived well into the eighteenth century. With its many fantastic attributes, the basilisk did not inspire confidence even to the imaginative seventeenth-century writers. Scientific biology had caught up with it long ago, its appearance was preposterous, no person had seen a live specimen for 100 years, and even its fabled generation from a cock's egg was doubted by many writers. The outbreaks of "basilisk hysteria" in the fifteenth and sixteenth centuries are likely to have been due to poisonings by filthy, unwholesome wells. The reason for the basilisk's puzzling longevity is likely that two veritable mainstays of its legendary attributes were founded in fact: birds resembling cocks can, in rare cases, lay eggs, and intestinal ascarids have been found inside eggs, masquerading as young basilisks. It is to be hoped that the frontline biological scientists of today shall be more favorably judged in 300 years than the sixteenth- and seventeenth-century savants who pondered the riddle of the basilisk. Superstition, exaggerated confidence in earlier authorities, and misinterpretation of factual observations in practical ornithology led them through an unavailing chase after an imaginary monster.

Although the classical basilisk, with its cock's head and tail of a serpent, is today only to be encountered in heraldry, its name—although not its distinguished position as King of Serpents—has been usurped by a family of large iguana lizards living in the rain forests of central America; their generic name is *Basiliscus*. The males of these modern basilisks have a high crest on their heads, sometimes resembling a crown or crest. They are beautiful, slender lizards that climb trees with great agility; the largest among them, the *Basiliscus americanus*, can grow to more than 3 feet in length. Although they cannot poison wells, break stones asunder with their eyesight, or peck holes in iron doors, the basilisk lizards have another remarkable accomplishment—one that is definitely not associated with the handiworks of the devil: they can walk, or rather run, on water, using their powerful hind legs and balancing their long tails. Some time ago, I had occasion to examine a young basilisk lizard. It demonstrated a remarkable quickness and a voracious appetite but lacked the sufficient run to make an attempt to walk on water. The basilisk studied its visitor with great interest, and under the penetrating stare of these reptilian eyes, one was reminded of Shelley's *Ode to Naples:*

> *Be thou like the imperial basilisk,*
> *Killing thy foes by unapparent wounds!*

Spontaneous Generation

They say "At first to living things the Earth,
At her formation, gave spontaneous birth;
When youthful heat was through the glebe diffus'd,
Mankind, as well as insects, she produc'd;
That genial wombs by parent Chance were form'd,
Adapted to the soil, which, after warm'd
And cherish'd by the Sun's enlivening beam,
With human offsprings did in embryo teem."

Richard Blackmore, *The Creation: A Philosophical Poem in Seven Books* (1712)

ARISTOTLE SPENT MUCH TIME PONDERING THE question of how all animals, plants, and other objects in nature were constructed, and how they had been engendered in the first place. He established an elaborate philosophical system that considered both the origins of life and the everyday formation of new individuals. Everything in nature, whether it belonged to the animal, plant, or mineral kingdoms, was made up partly of matter and partly of a form, or soul. Matter is the empty substrate, itself without capacity to exist; the form is what raises matter to a bodily shape or changes its characteristics. A block of stone thus has a form that makes it just that; when it is sculpted into a statue, it begets a new and higher form that incorporates the artist's intention. The form of a living creature, which could also be called a soul, was of course much more complicated than that of a block of stone, but its fundamental prerequisites were the same.

A key question for Aristotle and his adherents was how all the newly generated animals and plants were supplied with forms. It was well known that that humans and higher animals were bred by normal reproduction; their souls were thus engendered from those of their parents. According to Aristotle's *History of Animals,* certain lower animals "are not produced from animals at all, but arise spontaneously: some are produced out of the dew which falls on foliage . . . others are produced in putrefying mud and dung, others in wood, green or dry, others in residues, whether voided residues or residues still within the living animal." Eels and little fishes were spontaneously generated from the foam of the sea, insects from specks of sand whirled into the air by hot gusts of wind, and wasps from the putrid cadavers of dead horses. The old Egyptians believed that many different creatures—insects, frogs, mice, crocodiles, and numerous other strange beasts and reptiles—could be generated in the fertile humus left by the flooding of the Nile. Edmund Spenser describes these in his *Fairie Queene:*

> *As when old father* Nilus *begins to swell,*
> *With timely pride aboue the* Aegyptian *vale,*
> *His fattie waues do fertile slime outwell*
> *And ouerflow each plaine and lowly dale:*
> *But when his later ebbe gins to auale,*
> *Huge heaps of mudd he leaues, wherein there breed*
> *Ten thousand kindes of creatures, partly male,*
> *And partly female of his fruitful seed;*
> *Such vgly monstrous shapes elsewhere may no man reed.*

Aristotle's theory of spontaneous generation was as influential as his other teachings in philosophy and natural history; it was accepted, with reverence, not only among his contemporaries, but well into modern times. Virgil and Pliny added some embellishments of their own: they attached to each insect a corresponding type of putrefied flesh, from which this breed of insect could be generated. Thus, bees were formed from the flesh of oxen, as eloquently explained in the fourth book of Virgil's *Georgica,* on the authority of "an Arcadian Bee-master." A calf is beaten to death, but its hide is kept intact; it is put among fresh cassia and thyme and broken branches. The bee master then bides his time:

Meanwhile, within the marrowy bones of the calf, the humors
Grow warm, ferment, till appear creatures miraculous—
Limbless at first, but soon they fidget, their wings vibrate,
And more, more they sip, they drink the delicate air:
At last they come pouring out, like a shower from summer clouds,
Or thick and fast as arrows.

A lesser poet, Hieronymus Vida, the author of a lengthy poem on silk worms, improved on Virgil's idea. If a heifer was kept for 10 days without any water and fed only mulberry twigs and leaves and ". . . no other Meat or Drink; / Then kill and let her Carkass Rot and stink / When lo! a Lusty swarme of Silk-worms smal / Out of her Sides and Back you'll see to crawle."

Wasps and hornets were believed to breed from the carcasses of horses, and beetles from those of asses. Plutarch added that small rats could be formed from earth, and serpents from human corpses. That oysters were spontaneously generated no one doubted, least of all Opianus, the author of a poem on fishes: "They couple not nor bring forth any young / But of themselves they breed the Mud among / As Oisters do, Glib Oisters that no Male / Nor Female have." Ovid's *Metamorphoses* contains a beautiful section extolling the wonders of spontaneous generation, here presented, like many of the other classical poems quoted, as "Englished" by the seventeenth-century translator of Daniel Sennert's work on spontaneous generation:

So, when the seven channeld Nile forsakes the Plain,
When ancient Bounds retiring Streams contain,
And late-left Slime aethereal Fervors burn
Then various Creatures with the Gleab up turn:
Of those some in their very time of Birth;
Some Lame, and others half alive, half Earth.

Some authors believed not only that plants and animals could be spontaneously generated but that the origin of humankind was due to the parentless generation of the first individuals. Thus Virgil, in the *Aeneid,* spoke of men born from trunks of trees and stubborn oak, while Juvenal praised the virtues of the prehistoric "men born of the riven oak, or formed of dust," who had "no parents of their own." Pythagoras advocated a strange variation of these notions: he forbade the eating of beans, since these shared a common origin with the first humans! He imagined the beginning of life to have been a vast cesspool of corruption, in which the spontaneously generated animals and plants were born and sprang up from the earth. From exactly the same kind of putrefaction, Pythagoras claimed, men were formed and beans grew.

The traditional concept of spontaneous generation reigned supreme throughout the Middle Ages. Several early Arab naturalists of eminence, Averroes and Avicenna among them, accepted Aristotle's teachings about the parentless generation of lower animals. Avicenna even added that humans were by no means exempt from this general physiological law: his unprepossessing view of the origin of postdiluvian man was that they had been spontaneously generated from putrefied carcasses left over by the flood. Other medieval writers added their own touches: the twelfth-century Abbess Hildegard of Bingen maintained that reptiles had been generated from the ground stained with Abel's blood, and certain influential astrologers, Henry of Hesse and Michael Savonarola among them, emphasized the power of the stars in controlling the spontaneous generation of frogs, bats, mice, and fish. Leonardo da Vinci was yet another prominent supporter of this latter theory. The great Paracelsus, alias Theophrastus Bombastus von Hohenheim, merged the dogma of spontaneous generation with his own system of alchemy. Under the influence of the sun and the stars, the four elements could merge together and form most kinds of animals. Snakes, tortoises, and insects could be generated from putrid matter, and basilisks from the blood of menstruating women. Paracelsus even suggested experiments to artificially generate a small human being, a *homunculus,* from putrid excrements and semen; this foul-smelling matter was to be incubated in a glass flask and fed broth and nutritious fluids.

Another reckless adherent of Aristotle's theory of spontaneous generation was the influential Italian mathematician, naturalist, and physician Girolamo Cardano. Cardano was a universal genius in the manner of his time: he has achieved everlasting fame by being the first to publish a solution to the general third-degree equation. This was the most important sixteenth-century contribution to algebra. Another Italian mathematician, Nicolo Tartaglia, challenged his right to this discovery, however, and claimed that *he* had solved the equation and told Cardano how it was done, under the pledge of secrecy. Cardano's suspected breach of trust led to a lengthy feud, which was only one of many quarrels of his turbulent life. Some historians of mathematics have accepted Tartaglia's claim, but it is certain that Cardano was one of the most talented mathematicians of his time, and he may well have figured out the solution independently. His extensive biological writings are not always characterized by the same stringency as are his early mathematical flashes of genius, however. His opponent Herman van Booerhave astutely claimed that when Cardano was right, he surpassed all contemporary scholars in wisdom and farsightedness, but when he advocated erroneous theories, he surpassed them all in folly. A large section of Cardano's treatise *de Subtilitate* is devoted to various aspects of spontaneous generation. According to Cardano, not only lower animals, such as ants, flies, fish and crayfish, could be generated from putrid matter; many mammals, such as mice, hares, and gazelles, were regularly spontaneously engendered in this way.

In the early seventeenth century, the existence of a spontaneously generating force in nature was generally accepted. From this time onward, the terms *spontaneous generation* and *equivocal generation* became synonymous: the latter term emphasized the difference between the random, obscure process involved in the formation of lower animals from putrid matter and the straightforward, unequivocal generation of vertebrates. When the theologist Wolfgang Franzius published his *Historia Animalium Sacra* in 1612, claiming that no animal could be generated without parents, he was ridiculed by his contemporaries as being completely out of touch with modern biology.

A new era in the annals of spontaneous generation began in 1636, when the Wittenberg physician and philosopher Daniel Sennert published his *Hypomnemata Physica.* A section of this work was devoted to formulating a novel hypothesis of spontaneous generation, which differed from those of Aristotle and Cardano in several respects. Sennert was an early atomist, who refused to accept that the atoms of matter could independently rearrange themselves into a living creature with a soul. Since God was no longer capable of creating any new souls — not even that of a gnat — the spontaneously generated animals had to be derived from matter that had once been living. The soul of a horse was capacious enough to suffice for all the thousands of bees that were generated from its putrid cadaver. Everywhere in nature were forms, or souls, of plants, insects, and various spontaneously generated small animals, awaiting the opportunity to generate a living organism from putrid meat and offal. The novelty of Daniel Sennert's teachings, which later philosophers have called *seminalism,* is that he was the first to clearly deny the equivocal generation of living creatures from nonliving inorganic substances; new life could only develop from once living matter.

Similar notions had been advocated by Fortunio Liceti in his treatise *de Spontaneo Viventium Orto,* a solid tome dedicated entirely to all aspects of equivocal generation.

The Swedish Professor Olaus Rudbeckius the Younger was, like many late–seventeenth century biologists, a firm adherent of Sennert. He had once been consulted by a wealthy Swedish merchant, who asked him to explain a most puzzling experience. Before traveling to Stockholm, the merchant's wife had given him a box of pancakes made from duck's eggs, to be eaten on his way there. He forgot about them, however, and did not open the box until 8 days later, optimistically hoping that they were still fit to be consumed. To his extreme terror, numerous small frogs leapt out of the box as soon as it was opened; of the pancakes, no trace remained. While the skeptic would suspect that the merchant's wife, or perhaps one of his traveling companions, had eaten the pancakes and put the frogs in the box as a practical joke, Professor Rudbeckius had another, more learned explanation. He authoritatively told the merchant, who probably marveled at the advances of modern biology, that the ducks, from whose eggs the pancakes had been made, had probably been feeding on small frogs just before the eggs had been laid. The frogs' yet undigested "life seeds" were thus incorporated into the eggs. To an advocate of spontaneous generation, it seemed a natural thing that frogs could develop from these pancakes if they were allowed to ferment in an warm box for 8 days.

The great philosopher René Descartes wholly disagreed with Daniel Sennert: he denied that animals or vegetables possessed any kind of form or soul: they were just automata, without any higher qualities. A St. Bernard dog was of course a more accomplished creature than a tulip, but only in the same sense that a church organ had a more complicated mechanism than a beer barrel. According to Descartes, it took so little to make an animal that it was by no means surprising that worms and insects could be spontaneously generated from putrefying matter. Just as cheese could be made from milk, worms and larvae could be made from decaying cheese; matter was simply transubstantiated from one condition to another.

Another remarkable theory of equivocal generation, which seems like a mixture of the alchemy of Paracelsus and the seminalism of Sennert, was advocated by the German Jesuit Athanasius Kircher. In his monumental *Mundus Subterraneus,* published in 1665, he praised God as the great originator of equivocal generation. In the chaotic protomass, God had deposited a universal life force consisting of salt, sulphur, and mercury. When the fumes of this vital force came into contact with putrefying animals or vegetables, a large variety of beings could be spontaneously generated. Father Kircher described a number of simple experiments that he recommended the reader to try out himself, if he was unwise enough to doubt the existence of this vital force in all nature. If one wanted to create frogs, for example, one could collect some clay from a ditch where frogs were known to have been living and incubate it in a large vessel, while continuously adding rain water. Another popular experiment was to incubate a putrid lump of meat in an open vessel, to see the crawling maggots being spontaneously generated therein.

The Dutch naturalist Jan Baptista van Helmont described a spontaneous generation experiment that would be much more difficult to reproduce: he claimed that if wheat was incubated with water in a large flask covered by the skirt of an "unclean woman," a small living mouse would be generated within 21 days! The aforementioned Professor Olaus Rudbeckius had another odd notion: if an ounce of silver was dissolved in 3 ounces of nitric acid, with addition of water and 3 ounces of mercury, a little tree was spontaneously generated and grew on its own accord. This phenomenon is likely to have been a deposition of metallic matter occurring during the experiment, however. If Professor Rudbeckius had tried to expand the botanic garden of the University of Uppsala by using the repository of metallic solutions in the chemical laboratory, the harvest reaped from his strange alchemic "garden center" is unlikely to have been impressive.

In sixteenth- and seventeenth-century Britain, the belief in spontaneous generation reigned supreme. Even Sir Francis Bacon accepted the equivocal generation of insects, snails, eels, frogs, and toads. William Harvey, one of the greatest physiologists of the seventeenth century, was another supporter. Although he claimed, in his *De Generatione Animalium,* that every living thing must have been developed from an egg, his concept of an "egg" was very much different from that of modern biologists: it was an imperfect, often indeterminate object, the "first Rudiment from which an Animal doth spring." For example, Harvey considered a caterpillar as the egg from which the perfect imago developed by metamorphosis. Nor did he rule out that this "egg" could be spontaneously generated. He basically agreed with the current division of animals into the viviparous, oviparous, and spontaneously generated, writing that "some Animals are born of their own Accord, or (as they commonly say) of Putrefaction." Indeed, the vast majority of seventeenth-century British naturalists, including men such as Robert Hooke, Robert Lovell, and John Jonstone, were adherents to this ancient biogenetic fallacy.

When Sir Thomas Browne, who was otherwise another supporter of equivocal generation, dared to doubt whether mice could really be formed in this way, he was taken to task by an irate contemporary, the conservative Aristotelian Alexander Ross, in no uncertain terms:

> He doubts *whether mice can be procreated of putrefication!* So he may doubt whether in cheese and timber worms are generated; or if butterflies, locusts, grass hoppers, sel-fish, snails, eels, and such are procreated of putrefied matter . . . To doubt this is to question Reason, Sense, and Experience . . .

This widespread seventeenth-century belief in spontaneous generation was by no means confined to scientific works: the old fallacy was often referred to in contemporary literature. William Shakespeare several times alludes to the spontaneous generation of animals, most particularly in *Antony and Cleopatra,* where he speaks of serpents and crocodiles generated from the mud of the Nile. Similarly, Ben Jonson, in *The Alchemist,* speaks of bees, hornets, beetles, and wasps being generated from the carcasses and dung of creatures. In an amusing discourse written in 1592 (and just as valid today), Thomas Nashe compares the money-grubbing upstarts, adventurers made courtiers, and villainous "self-made" businessmen of the Elizabethan era with the imperfect, spontaneously generated animals. He likens "such obscure upstart gallants, as without desert or seruice, are raised from the plough to be checkmate with Princes" to the wretched insects bred without coition. "Brewers that, by retailing filthy *Thames* water, come in a few years to bee worth fortie or fiftie thousand pounds" are compared with foul creatures bred from corrupted water, and "Mother *Bunches* slimie ale, that hath made her and her fil-pot facultie so wealthy" likened to frogs bred from slime. Butchers getting rich by selling diseased and fly-blown beef are aptly compared with crawling maggots bred from putrid meat.

Jonathan Swift made an interesting allusion to spontaneous generation in *Gulliver's Travels.* When the tribe of noble horses, the Hoyhnhnms, debate whether to exterminate the subhuman Yahoos, one of them suggests that the first two of this abominable tribe suddenly appeared together on a mountain "whether produced by the Heat of the Sun upon corrupted Mud and Slime, or from the Ooze and Froth of the Sea, was never known." Dean Swift was well read in contemporary philosophy and biology and was evidently well aware of the musings of Paracelsus and Cesalpino that human beings could be spontaneously generated. The further meaning of Swift's passage about the spontaneous generation of the Yahoos has been debated, but in an admirable recent study, it has been clearly demonstrated that he used the abominable descent of the Yahoo race from mud and slime as an example of the brutish, uncivilized state of human beings without the Christian religion.

One of the most remarkable fallacies in the annals of spontaneous generation was the widespread credence that a class of so-called *zoophytes*, spontaneously generated hybrids between animals and plants, existed. Throughout the sixteenth and seventeenth centuries, it was vigorously debated whether a lamb could really develop from a plant, and whether young geese could grow on trees, hanging from the branches like ripe pears. These monstrosities were described in many zoological and botanical treatises of the time, and traces of the famous Vegetable Lamb of Tartary and the Barnacle Geese of Scotland still exist in the zoological terminology.

The Vegetable Lamb of Tartary

When, in the year 1356, Sir John Mandeville, a valiant and honorable knight residing in St. Albans, became incapacitated by advanced age and a troublesome rheumatic gout, he decided to write his memoirs. He had many things of interest to relate, since his life had been filled with travel and adventure. Mandeville's memoirs were considerably more interesting than those of a present-day senior civil servant or member of Parliament who sits down after retirement to depict his long and worthy career in the queen's service. As a young pilgrim, Sir John had set out for Palestine in 1322; in later years, his long and perilous journeys took him to Turkey; Great and Little Armenia; Tartary; Persia; Syria; Arabia; Upper and Lower Egypt; Libya; a great part of Ethiopia; Chaldaea; Amazonia; and Lesser, Greater, and Middle India. Sir John Mandeville's journeys combined the best of several worlds: like a time traveler, he was conversant with the hoary myths of antiquity, the marvels of the East, and the manifold wonders of medieval natural history.

Sir John had seen centaurs, hermaphrodites, cynocephalics, cyclops, pygmies, tailed men, men without heads, and "many other divers folks of divers natures." A certain Oriental tribe had enormous ears hanging down to their knees; another had upper lips of a prodigious size, which they used as parasols. He had seen the Dead Sea, in whose waters iron floated and feathers sank to the bottom, and the Mountain of Ararat, on the top of which the wreck of Noah's Ark could still be distinguished in clear weather conditions. He had visited the king of Calonak, who had more than one thousand wives, two hundred children and fourteen thousand tame elephants; the last were, in case of war, equipped with manned castles of wood to carry on their backs, to frighten the enemy host. At Polumbrium, Sir John had drunk from the Fountain of Youth; at Ceylon, he had seen a lake filled with the bitter tears shed by Adam and Eve after they had been banished from the Garden of Eden. Another far-famed marvel of the East was the gold-mining ants of Taprobane, which guarded their mines very closely; neither Mandeville nor any other traveler dared investigate the treasures of these ferocious insects further, since "deise Pissemyres ben grete as Houndes." In the river of Indus, Sir John Mandeville had seen 30-foot eels sport about; nor had he missed another zoological curiosity of these parts, the fearful giant rats of Thana. Only one of the marvels of the East had eluded the knight: about the Earthly Paradise he had nothing truthful to tell, since had never been there.

During his stay in Tartary, Sir John Mandeville was much impressed with the gluttony of the people: both the Grand Khan and his loyal subjects ate lion, dog, leopard, rat, and donkey meat with a voracious appetite, caring not whether their beefsteaks had been dressed or cooked in any way. A *boeuf Tartare* of leopard meat was likely to have been considered a rare feast even in fourteenth-century Tartary. But during his sojourn in Tartary, Sir John Mandeville had occasion to taste an even more singular local specialty. It was a large fruit, like a watermelon, that grew on a long stalk and, when opened, proved to contain a living little lamb. The Tartars greedily ate this strange being when fruits of this kind fell into their hands; with his first-hand knowledge of their dining habits, Sir John was unlikely to have been surprised by this intelligence. He was audacious enough to eat a piece of the lamb himself, probably served raw according to local custom. He does not describe what this vegetable lamb tasted like, but with the decorous piety manifested even in his tallest stories, he writes that "of that Frute I have eaten; alle thought it were wondirfulle, but that I knowe wel that God is marveyllous in all his Werkes."

Another version of the legend of the Vegetable Lamb of Tartary is given by Baron Sigismund von Heberstein, who had served as the German emperor's ambassador at the court of Muscovy, in 1517 and 1526. At this time, it was quite a dangerous position to be an ambassador in Moscow: the temperamental czars did not always honor the diplomatic privileges, and occasionally vented their rage about bad news from abroad on representatives of the foreign powers in question that were present in court. The czars were

A miniature illustration from a German manuscript of Sir John Mandeville's *Travels*, depicting the goose-tree and the lamb-tree growing side by side. Reproduced by permission of the British Library.

also sticklers about etiquette: an ambassador who forgot to remove his hat when meeting a Russian prince was punished by having the hat nailed to his skull by the palace guard. Baron von Heberstein was fortunate enough to end his diplomatic career without any untoward incidents. He was a man of learning and, after returning to Germany, published the *Rerum Muscovitarum Commentarii,* a valuable collection of historical anecdotes about Russia and its people. A courtier named Daniel Danielovich, described as a man of high veracity, had told him that his father was once sent to the Tartar king as an emissary from the Duke of Muscovy. In the neighborhood of the Caspian Sea, between the rivers Volga and Jaick, Danielovich the elder had seen a plant called the "Borametz" or "little lamb": it grew from a large seed the size of a melon and attained the height of about 2½ feet. Strangely to tell, it resembled a little lamb, with the eyes, ears, head, and wool, although it grew from a stem. It had hoofs, but they were not horny and rather resembled hairs brought together into the form of a cloven lamb's hoof. The living plant had blood, but its flesh rather resembled crab meat and was of so excellent a

An early illustration of the vegetable lamb, from Mandeville's *Travels*.

flavor that the defenseless plant was the favorite food of wolves and other rapacious animals. Another acquaintance of Baron von Heberstein's, a certain Michel, who served as an interpreter of the Turkish and Arabic languages to the Republic of Venice, had also heard of this plant. He claimed that its soft and delicate wool was often used by the Tartars as padding for the caps worn on their shaven heads.

Baron von Heberstein's description of the vegetable lamb was eagerly read by naturalists all over Europe. The living plant was considered one of the most remarkable examples of spontaneous generation: if a creature of this size could grow from a seed, it was not difficult to understand how simpler animals could be similarly generated. Girolamo Cardano was inspired by Heberstein's account to include a passage on the existence of living plants in his *De Rerum Natura,* published in 1557. He argued that if a being had blood it must have a heart, and that the soil in which a plant grows is not fitted to supply a heart with movement and vital heat. Nevertheless, he did not deny the vegetable lamb's existence, since if there were sponges in the sea, why should there not be similar plant-animals on land? Nor did he consider it impossible that, if the atmosphere was particularly thick and dense, a plant could have sensation and "imperfect flesh, such as that of mollusks and fishes." He knew for a fact that wolves ate the Boramez with avidity, although no other carnivorous animals were wont to touch it.

Cardano's greatest enemy, the megalomaniac Julius Caesar Scaliger, was always ready to attack him with cutting satire and insulting invectives. Scaliger's *Exotericarum exercitationum,* a solid tome of 1,200 pages, was wholly devoted to a heavy-handed criticism of Cardano. He had already become notorious through his fierce and scurrilous campaign against Erasmus of Rotterdam, which had disgusted many of his contemporaries. Scaliger was something of a skeptic, and Cardano's more imaginative musings in natural history—that the swan sings at the time of its death, that the bear forms its cub by licking, and that the peacock is deeply ashamed of its ugly legs—gave him ample scope for his vituperative talents. The section about the living plant was no exception. Scaliger chose to pretend that he was wholly convinced by Cardano's account of the Boramez lamb; he repeated all the ludicrous arguments and made them appear even more absurd. He ended his diatribe with an inquiry: "this is merely a little sauce and seasoning to your fable of the Lamb; but I would like to know from you how four distinct legs and their feet can be produced from one stem?"

The vast majority of the later commentators on the Boramez question seem to have missed the point of this heavy-handed irony altogether; they have misinterpreted Scaliger's intention to ridicule Cardano and instead proposed that Scaliger was just another believer in the vegetable lamb. After Erasmus died and an (unfounded) rumor of Cardano's death had begun to circulate, Scaliger wrote a hypocritical eulogy of these two men, cantingly bewailing that his incisive criticism of their works had led to the premature death of these two notable philosophers! Cardano recovered, however, and was by no means amused by Scaliger's peccadilloes; he replied to his criticism 2 years later, in no uncertain terms. He was given the satisfaction that Scaliger, who was the elder of these two quarrelsome Italian savants, predeceased him in 1558, mercifully oblivious that for 200 years, thesis writers all over Europe would claim that he had wholeheartedly agreed with Cardano on the subject of the Tartar lamb-plant. Girolamo Cardano lived until 1576, but his latter years were miserable indeed, since his children, for whom he had had great expectations, all met with unfortunate ends. His daughter was a pox-ridden whore, and his eldest son was a murderous, crippled hunchback who was finally executed for poisoning his own wife. His second son, who had made a career as a professional torturer, plotted to have his father executed by betraying him to the Inquisition, for having cast the horoscope of Christ. Although Cardano's old enemy Nicolo Tartaglia did his best to denigrate him, the wily old Cardano managed to elude this fiendish plot by appealing to one of his wealthy former patients, John Hamilton, the Archbishop of St. Andrews.

The early botanist Claude Duret was even more enthusiastic about the Boramez plant, which he described and figured in his *Histoire Admirable des Plantes.* Of all the trees, bushes, and herbs in the world, the vegetable lamb was the one that most filled him with pride and wonder before the creative power of God. It would have been more apposite to question why the Lord had bothered to create such a monster in the first place, since the lamb's life was by no means a pleasant one. The Boramez plant greedily devoured all bushes and herbs near its own stem, but when the crops within its limited reach had been consumed, the wretched creature succumbed to starvation. More often, the lamb died a bloody and violent death. According to legend, God had tried to place the defenseless plant under protection by forbidding all predators to attack it. The wolves did not honor this ancient agreement, however, and happily became vegetarians to have the opportunity to taste the tender flesh of the Boramez. The earthbound lamb could do nothing except utter a terrified bleat when surrounded by a pack of slavering wolves coming in for the kill.

According to another obscure old legend about the Boramez, it was one of the first plants created by God, and one of them grew in the Garden of Eden. It is unknown whether the figure of the vegetable lamb had a strange, religious significance or any connection with the early myths of Christianity. There is no direct evidence that this was the case, but the lamb is, after all, the symbol of Christ, whose body could, according to the traditional account of the transubstantiation, be miraculously resurrected from vegetable products: bread and wine. When the famous British botanist John Parkinson published his *Paradisus Terrestris* in 1629, its beautiful frontispiece depicted Adam and Eve wandering in the Garden, admiring the multitude of plants. In the background, a Boramez lamb is growing in one of the borders. John Parkinson had probably been inspired by the French poet Guillaume de Saluste du Bartas's famous poem *La Semaine,* originally published in 1579 and

A remarkable drawing of the Tartar lamb, from Claude Duret's *Histoire Admirable des Plantes*.

later skillfully "Englished" by Josiah Sylvester. This beautiful old poem describes how Adam takes a stroll among the newly created trees and plants in the Garden of Eden, which du Bartas seems to have imagined to have been planned rather like the Hampton Court maze:

> *Musing, anon through crooked walks he wanders,*
> *Round winding rings and intricate meanders.*
> *False-guided paths, doubtful, beguiling, strays,*
> *And right-wrong errors of an endless maze.*

He is startled to see a Boramez growing in one of the hedges

> *Feeding on grass, and th'airy moisture licking*
> *Such as those Barametz of Schythia bred*
> *Of slender seeds, and with green fodder fed;*
> *Although their bodies, noses, mouths and eyes,*
> *Of new-yeaned lambs have full the form and guise,*
> *And should be very lambs, save that for foot*
> *Within the ground they fix a living root*

Which at their navel grows; and dies that day
That they have browzed the neighboring grass away.

This part of the poem concludes with a beautiful eulogy to the marvelous lamb:

Oh! wondrous nature of God only good,
The beast hath root, the plant hath flesh and blood.
The nimble plant can turn it to and fro,
The nummed beast can neither stir nor goe,
The plant is leafless, branchless, void of fruit,
The beast is lustless, sexless, fireless, mute:
The plant with plants his hungry paunch doth feede,
Th' admired beast is sowen a slender seed.

Throughout the mid-1600s, the Boramez legend was still accepted by the scholars; in their weighty tomes, men such as Fortunio Liceti and Athanasius Kircher declared themselves convinced that the vegetable lamb existed, and that it was one of the most remarkable pieces of evidence of the reality of spontaneous generation. One of the few to object was Sir Francis Bacon, who was quite unconvinced that a plant could possess the power of thought and movement. He was supported by the German scholar Antonius Deusing, who pointed out that no living person had seen this marvelous lamb-plant, and that its supposed existence was a mockery to common sense and reason. His dissertation, which was published in 1659, might have convinced a few European scholars, but the Boramez plant was to remain a reality for many years to come. The Dutch merchant captain Jan de Struys reported a novel observation in an account of his travels to Russia and the Orient, published in 1681. He claimed that the Boramez was by no means an uncommon sight in the meadows on the western bank of the Volga. The Tartars were wont to harvest some of them, to make their soft, white fleeces into nightcaps, for the protection of their shaven heads from the night cold. They could easily be persuaded to sell the vegetable lambskins to travelers: Jan de Struys had several times purchased these skins and was always able to sell them for twice the money he had paid when back in Holland. One of these valuable vegetable lambskins was kept in the museum of the famous entomologist, Jan Swammerdam.

A coat lined with several beautiful Boramez skins was kept, for many years, in the Bodleian Library in Oxford. When, in 1570, the officer Sir

Richard Lea had been appointed Queen Elizabeth's ambassador to the court of the czar of Russia, he was determined to procure a Boramez skin. He wanted to bring it back to England as evidence that the vegetable lamb really existed, and also because he considered these skins as powerful antidotes against the plague "and other Noysome diseases." Several noble Russians assured him that there did "grow out of the ground certain living creatures in the shape of Lambs, bearing Wooll upon them, very like to the Lambs of England." A stalk like that of an artichoke grew out of the earth; it had a bud on it, which grew into the shape of a lamb. The Boramez ate the grass surrounding it, but "when it had eaten up the grass within its reach it would dye." The Russians assured Sir Richard that the Tartars were wont to flay the skins off these monstrous lambs and sell them, but they pointed out that these vegetable lambskins were very rare and valuable. But after Sir Richard had ordered a mortar to be sculpted from a large, beautiful agate that he had brought with him for use as trade goods, the czar himself declared himself willing to trade a coat lined with Boramez skins for this mortar. On his way back to London, Sir Richard Lea visited the court of the king of Sweden. According to a document entitled "Mr. Edward Smith's Relation of the Tartar Lambskin Garment given to the Bodleian Library in Oxford by Sir Richard Lea," which is kept in the archives of the Royal Society of London, King Johan III's first question to the English Ambassador was whether he had seen the skin of a vegetable lamb. When Sir Richard replied that he had not only seen but was actually in possession of several of them, the king envied him greatly. Johan III had himself, for several years, diligently been trying to procure a skin of the vegetable lamb but had never obtained any, despite the persuasive powers of the royal money coffers. Sir Richard was unwilling to sell him the lambskin coat, which he valued far above the "divers other rich Furrs, and other Rarities of great price" in his baggage.

Sir Richard Lea was very proud of his Tartar lambskin coat. A kinsman of his wife, Mr. Edward Smith, several times saw it in his house, and heard the remarkable story of how it was procured. Sir Thomas Bodley wanted to secure this valuable "toga ex lana agni Tartarici" for the Bodleian Library, and it was discussed in a report to the convocation on October 27, 1609. Bodley even had hopes of exhibiting the lambskin coat before King James, who had expressed an interest to see it. But Sir Richard was unwilling to donate it, for some reason, and it was not until December 2, 1615, that the Tartar lambskin coat was taken to Oxford, probably after the demise of its original owner. It was kept in a little chest or "Russian boxe." If Sir Richard had expected the Oxford dons to appreciate the rarity of this garment, his hopes were in vain. When, in 1624, his kinsman Edward Smith, a Councellor of the Temple, was visiting the Bodleian Library, he was shocked to find that the Tartar lambskin coat was hanging in Sir Thomas Bodley's own closet, and that it showed evident signs of wear and tear! Neither of the dons and librarians knew the rarity and worth of this garment, and Mr. Smith took them to task in no uncertain terms, in a letter to Mr. Rouse, the keeper of the library. According to the books of account of the Bodleian Library, a glover was paid 5 shillings to mend the lambskin coat in 1634. Ten years later, further expenditure was incurred for "sewing, mending and ayring the Tartar lambes coat." It is not unlikely that it later passed into the Oxford cabinet of curiosities collected by John Tradescant, since the catalogue of the first contents of his museum, published in 1656, mentions a "coat lyned with *Agnus Scythicus*." Its later whereabouts are uncertain, but since the Tartar lambskin coat is stated to be increasingly moth eaten for each stocktaking of the library, it is likely to have been devoured by insects near the end of the seventeenth century. Interestingly, the books of account of the Bodleian Library mention that a copy of the famous Boramez illustration in Claude Duret's book was painted and framed in 1643, probably for exhibition alongside the lambskin coat. According to the *Annals of the Bodleian Library,* this painting was kept in the Ashmolean Museum as late as 1890, but it cannot be traced today.

Another Scandinavian coveting a Boramez skin was Professor Olaus Wormius, of Copenhagen. When his friend, the Reverend Christian Stougaard, accompanied Prince Valdemar Christian on a journey to Russia, he brought back a fleece of the vegetable lamb as a present for Professor Wormius. It was kept at his famous *Musaeum Wormianum,* along with an elaborate painting of a Boramez lamb. The museum was later purchased by the king of Denmark, but, according to the still extant catalogues, both fleece and painting were lost in the late eighteenth century. Quite a few vegetable lambskins seem to have been kept at various museums in the mid-1600s, but none of them appear to remain today. The seventeenth-century trade in Boramez skins to Western travelers was impeded by a report from the German surgeon Engelbert Kaempfer, who had accompanied a research expedition to Persia and Russia. For many months, he vainly sought for a Boramez, searching the meadows, hedges, and forests of Tartary without success. The inhabitants of these parts always told him that the vegetable lambs were growing just beyond the next forest, or that the fields where they grew were infested with dangerous brigands with a dislike to Westerners. The persistent German did not let himself be fobbed off, however, until he had scoured most of Tartary in his hunt for the elusive lamb. The Tartars wanted to sell Boramez fleeces to him, but it was apparent to Dr. Kaempfer that these were skins from unborn (Astrakhan) lambs, which had been dried and prepared. He strongly suspected that the Boramez specimens kept at various museums had the same origin, and that there was no such thing as a vegetable lamb.

It was not until 1698 that any complete specimen of the vegetable lamb was brought before European naturalists. That year, Sir Hans Sloane, the famous secretary of the Royal Society of London, demonstrated a strange object, which most resembles a woolly branch of wood with four legs, slightly more than a foot in length. It had arrived in a Chinese cabinet of instruments and curiosities and had originally belonged to a certain Mr. Buckley; he had purchased it from a clever Indian merchant who had assured him that he was buying a genuine Boramez. It was apparent to Sir Hans Sloane, however, that the specimen was made from the large rootstock of a fern, which rather resembled the body of an animal, being covered with "a Down of a dark yellowish Snuff-Colour, shining like Silk, some of it a quarter of an Inch long." The individual who had manufactured it had cut the roots from the rootstock and fashioned four of the stems of the fern's leaves to resemble the lamb's legs.

Much later, in 1725, Sir Hans was able to demonstrate another vegetable lamb before the Royal Society. It belonged to the German physician Johann Philip Breyn, of Danzig, who had been given it by a friend who had purchased the lamb in Russia as a genuine Boramez. Dr. Breyn was highly impressed with its animal-like appearance. Indeed, it is likely to be the most skillfully faked Boramez ever constructed, although it has the air of a dapper little fox terrier rather than that of a stolid lamb; a tail and a dog collar are all that is needed to perfect its likeness to a vegetable dog! Dr. Breyn cherished his vegetable lamb, although he realized that it was a manufactured specimen, made from the rootstock of a fern, just like that described by Sir Hans Sloane 27 years In a Latin article, published in the *Philosophical Transactions* of the Royal Society, he ridiculed the credulous scholars and unreliable travelers who were responsible for the wide dissemination of the Boramez lore. None of them had seen the marvelous plant, but had been tricked by the skins and the faked lambs.

Dr. Breyn's elegant vegetable lamb, as depicted in the *Philosophical Transactions* of 1725.

The commerce with these odd tourist souvenirs seems to have been quite extensive: they were sold by Russians, Indians, and Chinese, the last being particularly skillful in fabricating the rootstocks into little lambs or dogs. Sir Hans Sloane and Dr. Breyn were proved right when the first of these large arborescent ferns was taken to Europe. From its woolly rootstock, the long, straight stems arose; these ferns could be more than 12 feet high, the large leaves included. Carl Linnaeus, who himself had seen a faked vegetable lamb taken from China to Sweden by a traveler, named this species of fern *Polypodium barometz.* In the mid-1800s, some of these ferns were taken to England, where they were successfully cultivated in the Botanic Gardens at Kew by Sir William Hooker; he gave them their present name: *Cibotium borametz.*

Unfortunately, both Sir Hans Sloane's original, seventeenth-century specimen and Dr. Breyn's elegant vegetable lamb seem to have been lost; at least, there is no record of either after the mid-1700s. When Britain's foremost connoisseur of vegetable lambs, Mr. Henry Lee, searched for Sir Hans Sloane's lamb in the 1880s, he was disappointed to find that both the vegetable lamb and the Chinese cabinet of surgical instruments and appliances in which it had arrived had been lost or mislaid long ago. Mr. Lee claimed that another lamb was kept at the museum of the Royal College of Surgeons at this time of his investigation. Although this statement has been doubted by some historians of the museum, it is hard to completely disprove, since the vegetable lamb might well have been yet another casualty of the horrific German bombing raid in May 1941, along with Chunee's bones, Frank Buckland's mermaid, and many other specimens.

Although the number of old Boramez specimens has been sadly diminished through the ravages of time, the venerable lambs are not yet a completely extinct species. A Boramez lamb, made from the rootstock and stems of a large fern, is still kept at the Natural History Museum in South Kensington. It resides in a large old chest of drawers, which has been its home for more than 200 years. This lamb was figured by John and Andrew Rymsdyk in their *Museum Britannicum,* published in 1778, as one of the most curious objects in the British Museum. I have had the opportunity to examine this lamb closely; it is not very well constructed, compared to Dr. Breyn's lamb, but is evidently very old. It is not, as judged by the contemporary illustration and description, the lamb that was shown to the Royal Society by Sir Hans Sloane in 1697. Its legs are straight, while Sir Hans's Chinese specimen was severely bandy-legged and not even able to stand up without assistance. Mr. Henry Lee, who examined this lamb in the 1880s, had no idea of its origins, but a study of Sir Hans Sloane's museum catalogues, which are also kept at the Natural History Museum, has convinced me that this great collector in fact had *two* vegetable lambs; it is the second of these, which he procured in the 1710s or early 1720s, that has been kept for posterity.

Surprisingly, it has turned out that yet another, much more handsome vegetable lamb has also withstood the ravages of time. Instead of being stored in a chest of drawers like its cousin at the Natural History Museum, it is one of the most prized exhibits of the Museum of Garden History in Lambeth. This specimen is really quite lamb-like and makes a strange

The vegetable lamb at the Museum of Garden History, reproduced with kind permission of the museum.

impression on the museum visitors, gazing at them from under its glass cupola. The style of arrangement of its show case is suggestive of the mid-1800s. It was formerly in the possession of a Cambridgeshire physician, whose family had owned it for more than 150 years. The only thing to mar the dignity of this venerable lamb is that it raises one hind leg, rather like a dog relieving itself.

Thus, not fewer than two vegetable lambs are to be found among the manifold natural curiosities of London; I am probably the only person to have patted them both on the back, although withstanding the temptation to take a bite from any of them, like Sir John Mandeville. Both lambs are probably quite old, dating back to the eighteenth century. Perhaps one of them was examined by Charles Darwin's grandfather, Dr. Erasmus Darwin, who was inspired to incorporate some stanza about the Boramez legend in his *Botanic Garden,* published in 1781. The "golden hair" of the lamb is clearly an attempt to assimilate the legend to the specimens constructed from fern roots. Strangely enough, Darwin moved its supposed place of growth from Tartary to the Arctic, where the soil would hardly be rich enough to nurture a plant of this kind:

Cradled in snow, and fanned by Arctic air,
Shines, gentle Borametz, thy golden hair;
Rooted in earth, each cloven foot descends,
And round and round her flexile neck she bends,
Crops the grey coral moss, and hoary thyme,
Or laps with rosy tongue the melting rime;
Eyes with mute tenderness her distant dam,
And seems to bleat—'a vegetable lamb.'

In 1887, the former naturalist to the Brighton Aquarium, Mr. Henry Lee, published a slender volume, today quite rare, about the Vegetable Lamb of Tartary. It was the product of many years of research and remains by far the best account of the lamb-plant and its lore. Henry Lee disregarded Dr. Breyn's notions that the faked lambs were the heart of the matter, since the vegetable lamb legend had existed for several hundred years before these specimens were exhibited. Instead, he proposed that the earliest versions of the legend, in which it was emphasized that the lamb was visible only when the fruit of the plant had opened, spoke in favor of the "vegetable lamb" in fact being an early misconception of the cotton plant. This theory has been repeated by several later writers on the subject, but there is good reason to doubt Mr. Lee's attempt at explanation. After the campaigns of Alexander the Great, the cotton plant had become well known throughout the classical world, but the Greek and Roman annals do not support the hypothesis of any strange myths linking it with vegetable lambs. This plant was also well-known to the old Chinese. It is hard to understand how such a strange misinterpretation of the cotton plant could have arisen, long after this plant and its practical usefulness had become common knowledge. Instead, there is reason to consider an old Chinese myth, first described by the German scholar Dr. Gustav Schlegel before the Eighth International Congress of Orientalists in 1889. Dr. Schlegel was an expert in translating old Chinese manuscripts and had found some startling new data about vegetable lambs. In a manuscript from the Tang dynasty, written in the mid-tenth century, some extraordinary methods in Chinese sheep farming were described. The shepherds waited until the lamb-plants started to grow and, when they were about to sprout, built a high wall to protect the plants from wild beasts coming to devour them. When the lambs were fully grown, the shepherds pulled down the wall, donned cuirasses and colorful robes, and galloped toward the wretched lambs on horseback, beating on drums and clashing cymbals to frighten them. The vegetable lambs, whose stems were connected with the earth, were terrified by this attacking horde and tore loose from their moorings, rending their navels with a shriek and running before the shepherds' horses to the waterpasturages. It was particularly emphasized that the lamb had to be scared out of its wits to tear loose from the stem; if an attempt was made to cut it off, the vegetable lamb died from the hemorrhage.

Dr. Schlegel encountered a similar rumor account of "lambs which grow simultaneously out of the ground" in a manuscript compiled prior to A.D. 429, speaking in favor of the vegetable lamb legend being very old indeed. The vegetable lamb legend is also mentioned in a manuscript written in the middle of the eleventh century, and in the undated statement of a certain Chang Ye about his voyage to the West. Interestingly, a rather similar legend was extant among the old Jews. An old Hebrew book entitled the *Talmud Ierosolimitanum,* written by a certain Rabbi Jochanan in the year 436, tells of a strange plant-animal called the Jeduah. Its form was like that of a lamb, but it grew from a stem. When the hunters fired arrows through this stem, the creature fell to the earth and died. Its bones, put into the mouth of a soothsayer, endowed him with the gift of prophecy. According to another version, the Jeduah had a human shape and was most ferocious: it attacked every creature unwise enough to reach within its limited boundary. When it had devoured all the foliage within the tether of its stem, this vegetable, semihuman being died; more commonly, it was killed by hunters who cut or shot off its stem, whereupon the creature expired.

It can thus be proven that a Chinese and Jewish tradition of vegetable lambs similar to the classical European version, as told to Sigismund von Heberstein in 1526, existed already in the fifth century A.D. This version particularly pointed out that the lamb grew from a powerful stem connected with its navel and did not develop in large fruits—a claim that makes Henry Lee's hypothesis extremely unlikely. Similarly, Gustav Schlegel's own theory, that the vegetable lamb was really the camel, can be disregarded as absurd and in complete disagreement with the known facts about the Boramez legend. A later hypothesis, published by Dr. Berthold Laufer in 1915, that the vegetable lamb was a misconception of the thready excrescences of certain seashells, seems equally unconvincing. Instead, the vegetable lamb legend seems to have been taken over from a preexisting Chinese tradition that existed already in the fifth century, and probably even earlier; it is impossible to state the definite origin of this tradition, but it is unlikely to have been inspired by either the cotton plant, the camel, or the pinna seashells. The legend is likely to have been spread from China to other parts of the Orient, such as India and Tartary, and was first encountered by Western explorers in the fourteenth century. It is more difficult to explain why the vegetable lamb myth became even more widespread in the seventeenth century, when the nonexistence of the vegetable "water-sheep" should have become evident even to the Chinese and the Tartars. Some sources indicate that the Tartars and Indians enjoyed to make fools of the expeditions of Western explorers who came to search for the Boramez plant, and that they deliberately fed them tall stories about the vegetable lamb and its strange habits. The incorporation of the Boramez into scientific zoology made them succeed beyond their wildest hopes. Well into the eighteenth century, scientific expeditions venturing into Tartary kept a regular lookout for the vegetable lamb, but without finding any trace of them other than the skins and faked lamb specimens sold to them by the inscrutable Orientals.

As late as 1791, the French botanist Dr. de la Croix followed Guillaume du Bartas and Erasmus Darwin in extolling the strange lamb-plant in a strange Latin poem, which was later translated by that erudite biographer of the vegetable lamb, Mr. Henry Lee. An explorer is on his way to Asia, when

. . . in his way he sees a monstrous birth,
The Borametz arises from the earth:
Upon a stalk is fixed a living brute,
A rooted plant bears quadruped for fruit,

It has a fleece, nor does it want for eyes,
And from its brows two woolly horns arise.
The rude and simple country people say
It is an animal that sleeps by day
And wakes by night, though rooted to the ground,
To feed on grass within its reach around.
The flavour of Ambrosia its flesh
Pervades; and the red nectar, rich and fresh,
Which vineyards of fair Burgundy produce
Is less delicious than its ruddy juice.

The Barnacle Geese

Sir John Mandeville's *Travels* were immensely popular among his contemporaries, and numerous medieval manuscripts of this early bestseller are still kept. I myself have seen an English and a German manuscript of the *Travels* at the British Library's Department of Manuscripts; both of them are beautifully written, with many illuminated and hand-colored miniature illustrations. After the advent of book printing, the gouty old Knight's adventures were published in a multitude of editions. The French original was translated into English, Latin, Spanish, German, Italian, Czech, Irish, and Danish, and Mandeville's *Travels* became one of the most commonly read books of the sixteenth and seventeenth centuries. More than 250 manuscripts of Mandeville's *Travels* exist, written in ten different languages, and there are about 180 printed editions known.

From the late 1700s onward, certain suspicious literary historians closely examined the sources of Sir John Mandeville's masterpiece: they found that it consisted almost entirely of unacknowledged extracts from other, earlier travelers, with certain imaginative additions from Mandeville's own pen. Throughout the nineteenth century, Mandeville's *Travels* were considered a worthless fraud by ethnologists and antiquaries, and their author was mocked as a mere plagiarist. Certain literary historians even suspected that there had never been any Sir John Mandeville, suggesting that his alleged reminiscences were instead the work of a Liège physician named Jean de Bourgogne, also known as Jean à la Barbe, or Bearded John. In our own time, Mandeville has achieved a well-deserved renaissance, however: although his memoirs are definitely a forgery, literary historians agree that they have been compiled in a particularly skillful manner and presented in a forceful and stylish prose.

The part of Mandeville's *Travels* that dealt with the Tartarian lamb was plagiarized from the reminiscences of the much-traveled Italian monk Odorico of Pordenone, the earliest European to visit Tibet. That he had eaten a piece of the lamb was Mandeville's own invention, however; the strict Franciscan monk would not have dared eat of such a monstrous being without the sanction of the pope. After the Tartars had presented the vegetable lamb, Friar Odorico told them about a similar wonder of nature that occurred in certain Scottish Isles: a particular kind of geese grew on trees and hung by their beaks from the branches of the goose-trees until mature and ready to fly away. In an unique medieval manuscript called the *Livre des Merveilles,* kept at the Bibliothèque Nationale in Paris, this scene is symbolized by a beautiful miniature painting. The turbaned Orientals solemnly hand a vegetable lamb over to Friar Odorico and his companions, and the cowled friars give them, in return, a branch of the Scottish goose-tree, with the young birds still hanging by their beaks from the tree's branches. The East and the West meet here, for one of the first times, and exchange their greatest marvels.

The earliest origin of the Barnacle goose myth is, like that of the Vegetable Lamb of Tartary, unclear. The earliest authoritative account of these marvelous birds is that of Giraldus Cambriensis in 1187, but he is unlikely

A vegetable lamb and some tree-growing geese in a beautiful miniature from the manuscript *Livre des Merveilles,* written in 1433. Reproduced by permission, Cliché Bibliothèque nationale de France, Paris.

Tree-growing geese: a drawing from the Salisbury Bestiary, reproduced by permission of the British Library.

to have been the first to describe them. Although obscure, certain tenth- and eleventh-century sources may well be alluding to the tree-born geese. The barnacle geese were unknown to the ancients, and there is not a single reference to them in the traditional Greek and Roman sources. The hypothesis of the erudite biographer of the Barnacle geese, Mr. Edward Heron-Allen, F.R.S.—that certain figures in Mycenean decorative pottery represent convulsed-looking, "barnaculized" geese—is highly speculative and does not account for the massive silence of the classical writers on the subject of tree-born geese.

Nor is it known exactly where the barnacle geese myth originated. Arabian, and even Chinese, versions are known, but several sources point to Britain. The majority of early writers place the breeding place of the barnacle geese in Ireland or Scotland. A famous riddle in the eighth-century *Book of Exeter* asks:

In a narrow was my web, and beneath the wave I lived;
Underflowen by the flood; in the mountain-billows
Low was I besunken; in the sea I waxed
Over-covered with the waves, clinging with my body
To a wandering wood—.

Quick the life I had, when I from the clasping came
Of the billows, of the beam-wood, in my black array;
White in part were my pranked garments fair,
When the Lift upheaved me, me a living creature,
Wind from wave upblowing; and as wide as far
Bore me o'er the baths of seals — Say what's my name?

According to some philologists, the correct answer is — a barnacle goose.

The learned Giraldus Cambriensis, who had visited Ireland during the reign of Henry II, wrote in his *Topographia Hiberniae* that the country harbored a kind of geese called *Bernacae,* which were formed in a most extraordinary way. They were produced from logs of fir timber tossed along the sea. The geese hung down by their beaks like seaweed, and derived their nourishment from the sap of the wood and the froth of the sea. When very young, their bodies were surrounded with shells, but later, they grew a strong coat of feathers and were able to tear loose from the log and fly away. The barnacle geese were never seen to breed, nor to hatch eggs or build their nests, since new individuals were constantly spontaneously generated out at sea. Bishops and religious men in these parts of Ireland did not scruple to eat the spontaneously generated geese, even during Lent, arguing that they could not be counted as meat, since they were neither flesh nor born of flesh, but creatures bred from tree and water. A roasted goose, served with a bottle of Burgundy from the cellars, must have been a welcome respite from the tiresome fasting period. The strict Giraldus doubted whether this practice was really pleasing to God, however, and he made a most astute analogy. He compared the bishops' eating these geese to a gluttonous savage devouring one leg of Adam, our first parent. Although the First Man was not born of flesh, the cannibal in question could not be adjudged innocent of having eaten meat! Pope Innocent III agreed with him: at the Fourth General Lateran Council in 1215, he issued a bull forbidding the eating of barnacle geese during Lent; this must have led to much bitter murmuring and gnashing of teeth among the less ascetic monks.

Another version of the barnacle goose myth, which was widespread in the late Middle ages, told that certain trees near the Scottish coast bore fruits from which the geese were developed, hanging from the tree's branches by their beaks. The goose tree was always growing near the waterline, and its crown leaned out over the water, enabling the geese to plunge into the waves when mature. In his thirteenth-century bestiary, the Frenchman Pierre le

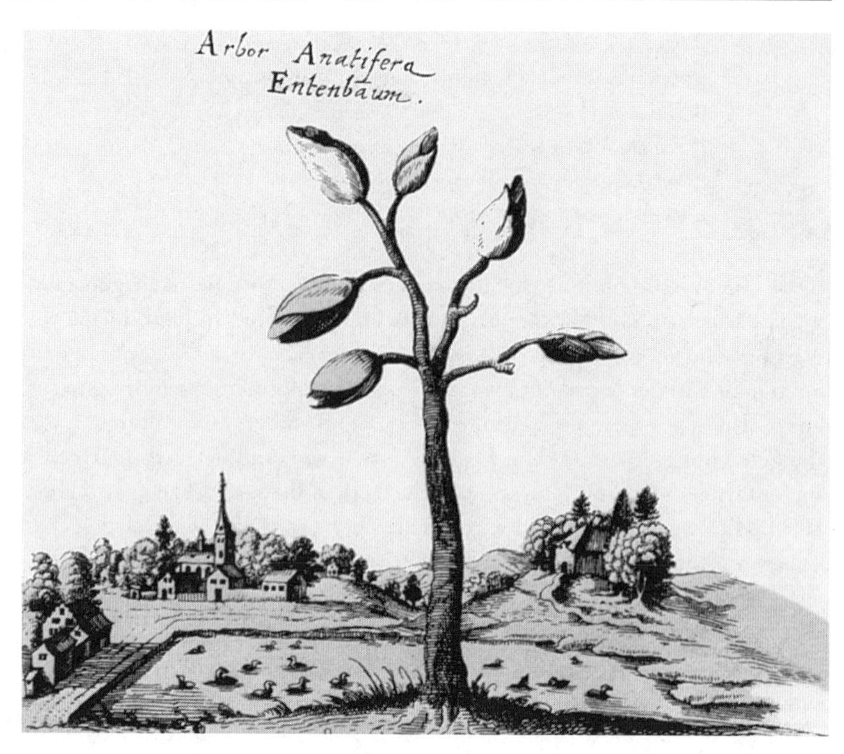

The goose-tree as drawn in the *Dendrographia* of John Jonstone.

Picard told another story: like pears, the birds dropped from the tree's branches when it was shaken by a storm. The birds that fell onto the shore were crushed to death, but those falling into the water were saved. In the usual moralizing tone of the bestiaries, this wonder of nature is given a parallel in human life: the people who are not washed in water (baptized) while young are as utterly lost as the crushed and broken birds on the stony shore, when they face God after the Last Judgment.

The majority of medieval writers placed the goose-bearing tree in Scotland, Ireland, or the Orkneys; some of them claimed that the spontaneously generated birds also occurred in Wales, England, and the Scandinavian islands. Gervase of Tilbury, writing in 1211, stated that great numbers of geese grew on the young willow trees that abounded near the Abbey of Faversham in Kent; these birds were eaten as "fish" during Lent. One of the earliest opponents to the barnacle goose myth was King Frederick II, who kept a large menagerie at his court in Sicily and wrote one of the earliest treatises on hawking. He considered this myth to be absurd, and presumed it to have arisen from ignorance about the actual nesting places of these birds. He was seconded by the learned Dominican Albertus Magnus, who was otherwise a firm supporter of spontaneous generation. In his *De Animalibus Historia,* written about 1250, Albertus Magnus declared that he and his friends had several times seen the so-called barnacle geese lay eggs and hatch them out in the ordinary way. Another clerical gentleman who did not just read about the marvelous birds in the works on natural history was Aeneas Sylvius Piccolomini, who was sent on a secret mission to Scotland in 1435. At the court of King James II, Piccolomini made inquiries about the true whereabouts of the celebrated Scottish goose-trees, perhaps hoping to bring a branch from one of them with him to Rome. The king and his entourage had little new to tell, but Piccolomini was undaunted. He traveled to the northern parts of Scotland, but without finding any barnacle geese; even then, the natives assured him that the goose-trees were growing in the Orkneys. Piccolomini, who later became pope under the name Pius II, later wisely commented on his wild goose chase in Scotland, with the words that wonders always recede further away when diligently sought after.

A new era in the annals of the barnacle geese began in 1527, when the Scottish historian Hector Boethius, who was a professor at the University of Paris, published his *Historia Scotorum,* which contained a section on the barnacle geese, or *Claiks,* as he called them. He was well aware of the legend that these birds were generated from the fruits of trees near the ocean, but he was inclined to disbelieve it; he had sailed through the seas where these wonders were said to occur without finding a single specimen of this remarkable tree. Instead, he had once seen a huge tree trunk, which had floated ashore near the castle of Petsligo in 1490, that was bored through by worms and had a quantity of barnacles hanging from it. When examined, these barnacles seemed to have heads, feet, and wings; others had feathers and seemed like perfectly shaped birds. When sawed through, the tree trunk was observed to be swarming with worms inside. People came from miles away to see this wonderful bird-tree, which was finally deposited in the Church of Saint Andrew, beside the town of Tyre. Hector Boethius and Master Alexander Galloway, Parson of Kinkell, both believed that the worms inside the log had been generated by "the nature of the Ocean sea," and that they later developed into barnacles, which in their turn became fully fledged birds.

All the sixteenth-century writers of cosmographies and encyclopedias on natural history reviewed the barnacle goose myth in detail. One of the most erudite accounts, illustrated with a beautiful woodcut of the goose-tree, was that of Ulysses Aldrovandi, in his *Ornithologia.* In the early seventeenth century, three versions of the old myth existed in parallel, causing some confusion among the naturalists. The early botanical writer Claude Duret, who had shown such enthusiasm before the vegetable lamb, dedicated an entire chapter in his *Histoire Admirable des Plantes* to describing the trees that were capable of breeding living creatures. He first reviewed the myth of the Scottish goose-tree, with the birds hanging from their beaks from the branches before plunging, when fully fledged, into a small pond that providence had placed just under this marvelous tree. In *La Semaine,* Guillaume de Saluste depicted poor Adam, who had only just recovered from the shock of seeing the vegetable lamb, encountering another strange botanical specimen:

A remarkable drawing of the goose-tree, from the *Ornithologia* of Ulysses Aldrovandi.

A tree whose fruits turn into fish if they fall into the water, and into birds if they fall on shore. From Claude Duret's *Histoire Admirable des Plantes*.

So slowe Boötes underneath him sees
In th'icy isles, those goslings hatched of trees,
Whose fruitful leaves, falling into the water,
Are turned (they say) to living fowls soon after;

Himself, Duret preferred another version of the myth: the one in which the fruits of the tree developed into birds first when they fell into the water. He also reviewed, and illustrated, an even taller tale: that certain trees in Ireland had fruits that developed into fowls if they fell on dry land, and into fishes if they fell into the sea. This marvelous tree was also praised by the eloquent Guillaume de Saluste:

. . . that Tree from off whose trembling top
Both swimming shoals, and flying troops do drop:
I mean the Tree now in Juterna *growing,*

Whose leaves disperst by Zephyr's *wanton-blowing,*
Are methamorphos'd both in form and matter;
On Land to Fowles, to Fishes in the Water.

The majority of seventeenth-century writers doubted the existence of an actual goose-tree, preferring instead the version of Hector Boethius, that worms generated from rotting timber could develop into barnacles, and later into geese. Several seafarers and zoologists had made similar observations, commenting on the bird-like aspect of the animal inside the barnacle's shells. King François I of France, who was interested in the barnacle geese, once ordered some naturalists to dissect a barnacle in his presence; the king and courtiers were amazed at the sight of the bird inside the shellfish:

So, rotten sides of broken ships do change
To barnacles, O transformation strange!
'Twas first a green tree; then a gallant hull;
Lastly a mushroom; then a flying gull.

During the mid-1600s, the barnacle goose myth was challenged by several writers. In 1596, a party of Dutch explorers, who were searching for the northwest passage to China, were nearly shipwrecked in a violent storm. After several days of perilously keeping afloat, their vessel was beached on an uninhabited, snowy island belonging to the desolate Novaja Zemilja archipelago. Here, they unexpectedly encountered a large colony of barnacle geese, which did not, however, act as described in the books of natural history. Instead of hanging by their beaks from the branches of the goose-tree or growing from waterlogged logs of wood, they were peacefully resting on their eggs, and their nests were scattered among the snowy ridges of the island. The geese were rudely awakened when the starving Dutchmen gave a great hallo and charged forth to seize a couple of birds. The barnacle geese flew up cackling "Rot! Rot!" and the Dutchmen managed only to tackle one of them, which they cooked and devoured, served with a huge omelet made from sixty of the birds' eggs. The German naturalist Antonius Deusing, who had been active in the campaign against the vegetable lamb, published a thesis about the barnacle geese in 1659, in which he doubted the traditional account of their origin. He referred to the Dutch travelers' tale, which had been published by Gerrit van Veer in 1609. Since it had now been proved that the birds in question nested and lay eggs, the old tradition about their miraculous generation up in the trees, or from barnacles adherent to rotting timber, seemed more ludicrous than ever. Deusing also queried why no one had ever been able to cultivate a branch of the goose-tree in a botanical garden, or even to bring a branch or log of it to be examined by men of observation. In Sir Joseph Banks's collection of theses on birds, kept at the British Library, can be found two other seventeenth-century works devoted to elucidating the riddle of the tree-born geese. In his thesis *Ex Physicis de Ortu Avis Britannicae,* defended at the University of Wittenberg in 1665, Dr. Johannes Ernestus Hering likewise refuted the opinion that these British geese were developed from the sea, from trees, from putrefaction, or from other matter, according to the marvelous ideas abounding at the time. In a later thesis, the *Discursus Historico-Physicus de Avis Britannicae,* Dr. Georgius Funccius agreed with his colleagues that the legend of the barnacle geese seemed wholly exaggerated.

These three skeptical German academics were answered by their countryman Athanasius Kircher, whose great appetite for curiosities had, as usual, got the better of whatever sound judgment he possessed. The imaginative Jesuit proposed that even though the geese laid their eggs on the icy ridges of Novaja Zemilja, these eggs fell into the water when the ice melted and then floated, like a row of miniature buoys, down to the Orkneys. Their contents stuck to rotten timber and later metamorphosed into complete birds. Another distinctly odd interpretation of the barnacle goose myth was proposed by Father Kircher's Jesuit colleague Father Philip Buonanni, in 1681. This clerical gentleman, who was apparently quite sane, believed that immense quantities of eggs of these birds were laid on the ice and later fell into the cold sea: enough eggs to nourish all Europe, if only they could be collected! When the ice melted in spring, the eggs broke and their contents were mixed with the water. After floating to the Orkneys, the Hebrides, Iceland, and Norway, these "scrambled eggs" developed into perfect birds after adhering to old logs.

The learned Dane, Professor Olaus Wormius, had several barnacles at his museum that were classified as young geese. A French colleague had told him that many barnacle geese, some driven on foot, others "salted" and brought in barrels, were taken from Normandy to Paris during Lent, to be eaten as shellfish by religious men. Some worldly Parisians fearing that their partiality to roasted goose might endanger their immortal souls, consulted theologians from Sorbonne University. The learned clerics replied that they were in no danger whatsoever of being banished to the nether regions, since these geese could well be considered as fish. Thomas Bartholin, the foremost pupil of Olaus Wormius, shared his interest in barnacle geese. From his friend Otto Sperling he had received a couple of barnacles, the innards of which he had examined closely with a microscope. It was soon clear to him that this creature was "an insect or a worm" rather than a bird embryo. Bartholin was thus the first to refute the barnacle goose myth through his own experimental observations. His accurate drawings of the opened barnacle were the earliest of their kind. Bartholin was an acknowledged authority on all aspects of natural history, and it is astounding to note that the barnacle geese myth survived his attack unscathed, even in his native Denmark.

In 1697, the Copenhagen physician Erich von Möinichen published a thesis entitled *Conchae Anatiferae Vindicatae*—The Duck-Breeding Shellfish Vindicated! After a turgid, lengthy discussion of the opinions of a vast number of earlier writers on this subject, he boldly declared himself a supporter of the dogma that geese could be formed from barnacles. Two years later, Thomas Bartholin's nephew Holger Jacobsen was more cautious: in a catalogue of the exhibits of the Royal Museum in Copenhagen, he described the barnacles of Olaus Wormius and recounted the barnacle goose legend at length. But he ended his discourse with the guarded comment: "but whether real geese are produced is a matter of comment among the more sagacious." If Erich von Möinichen or Holger Jacobsen had wanted a suitable epitaph on the barnacle geese, they could have quoted another Scandinavian, Bishop Haquin Spegel, whose eleven thousand–stanza poem on the Creation contains the following lines (here translated into English):

Somewhere in Scotland, a lake, or rather bog,
Is constantly surrounded by a cloudy fog
The trees around this lake astounding fruits do bear;
Since it is not a nut, an apple or a pear,
But little ducklings, many people say,
Are growing from their branches until fit to fly.
Once fully fledged, they plunge into the sea;
That this is true, all learned men agree.

It is amazing to contemplate that the absurd old legend of the barnacle geese was the longest-lived of the traditional beliefs in the spontaneous generation of vertebrates. Both in Scandinavia, Germany, and France, the barnacle geese found new friends well into the eighteenth century. In 1735, the historian Jonas Ramus declared himself convinced that geese were everyday spontaneously generated from wrecks and waterlogged trunks of trees near the Norwegian coast. Another Norwegian, Bishop Hans Egede, the author of a book on the geography and natural history of Greenland, considered the "treeborn ducks" one of the foremost natural curiosities of these parts. They were developed from a viscous substance that adhered to waterlogged trunks of wood. A little worm was first engendered and later developed further into a barnacle, which was transubstantiated into a fully fledged goose that pecked a hole in the barnacle's shell and flew out. A similarly absurd hypothesis was advocated by the Frenchman A. F. B. Deslandes: the goose pecked at the shell of the barnacle to break it open and pull the animal out, later laying its own egg between the shells! As late as 1757, his countryman Dezallier d'Argenville brought forth another ingenious hypothesis: that the goose forced open the shells of the barnacle and deposited its egg in there, like some bizarre marine cuckoo, leaving the wretched crustacean in charge of the care and nurture of its offspring.

The French debate about the barnacle geese was finally ended in the 1780s, by the zoologists Guéttard and de la Faille. M. Guéttard published a critical review of the many different descriptions of the tree-growing birds, to lay the ghost of the barnacle geese once and for all. He was amazed that there were still people who believed that barnacles could develop into young geese. De la Faille had cultivated barnacles on rotten logs of wood in an aquarium, without observing any signs that they would metamorphose into young birds. Like the majority of ornithologists, he was aware that there existed a particular kind of black geese that occurred in considerable flocks in autumn and winter, although they were never seen to breed or hatch their young. De la Faille was well aware, however, that the birds had normal genital organs; he presumed that they were hatching their goslings in some desolate Arctic territory. From the scholarly journals, this final debunking of the barnacle goose myth spread to the newspapers. The journalists were amazed that such a ludicrous old fallacy could have remained within scientific zoology for more than 800 years. Since this time, the expression "l'histoire d'un canard" has been used to deride implausible tall tales. The present-day expression *canard* is thus a direct descendant of the tree-borne Scottish geese. M. Guéttard recorded that, at his time of writing, the habit of eating barnacle geese during Lent still remained among the common people. In the British Isles, this ancient custom remained well into the twentieth century. In 1913, an Ulster clergyman wrote to inform the Zoological Society of London that his parishioners freely ate barnacle geese during the fasting period.

Only one of the barnacle specimens kept in various museums during the heyday of the barnacle geese seems to be extant today; it is kept at the Museum of Natural History at Bordeaux. The barnacles of Olaus Wormius have been lost, as have the "Barnacles, four sorts" that were classified as "Whole Birds" in John Tradescant's museum. The Dutch naturalist Adriaen Coenen had an entire barnacle goose, plucked, smoked, and salted, in his museum, but it was lost (perhaps eaten!) by marauding Spanish troops in 1578. The latest of these exhibits was the "Barnacle Tree, or tree bearing geese, found at sea by Capt. Bytheway," which was exhibited before the London public at Spring Gardens in 1807. Probably, it was an old log covered with barnacles; the captain's odd name arises suspicion that it was an elaborate hoax.

Carl Linnaeus was apparently well aware of the barnacle goose legend, which was still adhered to by certain conservative Scandinavians when he was a student in the 1720s. A rigid opponent of spontaneous generation, he was of course wholly unconvinced by the legend. In his system of zoology, he gave the tribe of sea geese characterized by their dark coloring and somewhat smaller build the name *Branta bernicla,* as a memorial to their supposed origin from barnacles. According to the philologist Max Müller, the word *barnacle* itself was derived from *hibernicula,* since the birds were originally considered to be natives of Ireland. Ornithologists now differentiate between two species of black geese: the Brent goose (*Branta bernicla* of Linnaeus) and the barnacle goose (*Branta leucopsis* of Becht). Neither of the geese breeds anywhere on the British Isles or the European continent, but they are winter visitors to these parts. The barnacle goose breeds in Spitzbergen and east Greenland, often on the faces of high cliffs; the young goslings, when leaving the nest, do not "plunge into the frothy sea," but have to scramble or fall 1,000 to 1,400 feet down perilous precipices, toward the unyielding, icy surface below. The Brent goose breeds in colonies on many locations in the Arctic, including Spitzbergen and the Novaja Zemilja. It is very likely that the hungry Dutch explorers, in 1596, had the good fortune to find one of these colonies. In German, these birds are still called Baum-Gans, or rather Rot-Gans; the latter name is due to their powerful cackling sound, which was also remarked on by the Dutch seafarers. The reason that the Brent and barnacle geese were accused of growing from trees or from old logs was that huge flocks of these birds came to Germany, Holland, and Britain every year. Since no person had ever seen them nest or hatch their young, it was widely presumed that the geese were spontaneously generated at sea. The statement of Albertus Magnus that he had seen barnacle geese nest in Germany cannot be uncritically accepted, since these birds today do not nest or breed anywhere near these parts. It cannot, of course, be excluded that the black geese were natives of Germany during this early period, but if this was really the case, it is inexplicable why other writers did not use this fact as an argument against the tale of their generation from trees or barnacles.

Linnaeus was also inspired by the barnacle goose legend when he named the barnacles in his system of zoology. He called these remarkable creatures, which belong to the subclass Cirripedia among the crustaceans, *Lepas anatiferae* (anatiferae = duck breeding). As a larva, the species leads an active and free-swimming life, developing through three stages and changing its appearance considerably. In its third stage, it develops its bivalve shells but still moves around vigorously, paddling with its six pairs of forked legs. The barnacle then selects the spot where it will spend the remainder of its existence, and presses its head against the surface. A kind of cement pours from its antennae, hardening at once in the water, and the barnacle is settled for life. It then grows rapidly: its shells are replaced with hard calcareous plates, and its legs become feather-like tentacles, or cirri. The witty zoologist Sir Ray Lankester likened the strange life cycle of the barnacle to the experience of an active, healthy schoolboy, running about with great vigor, who suddenly has his head super-glued to the pavement, thus being forced to spend his entire adult life upside down, kicking food into his mouth with his feet!

The End of Spontaneous Generation?

During the second half of the seventeenth century, the spontaneous generation controversy became increasingly heated. On one side were an unholy alliance of Paracelsists, atomists, Cartesians, and conservative Aristotelians; from vastly different standpoints, they all supported spontaneous generation. On the other side was a steadily growing number of experimental biologists who attempted to disprove spontaneous generation through their own observations. These biologists had support, however, from the religious traditionalists, who believed that God had once, at the beginning of time, created all animals and plants that were intended to inhabit the Earth. Since it was His intention that they were to be fruitful and propagate their species, He did not need any help in populating the Earth, least of all through the spontaneous generation of various foul creatures from putrid matter. It was blasphemy, they asserted, to suggest anything of the like.

One of the several prominent scientists who began to doubt the concept of spontaneous generation was the Florentine court physician Francesco Redi. In the 1660s, he planned an ambitious series of experiments to prove or disprove the equivocal generation of insects. He procured numerous carcasses of as many kinds of animals as possible: snakes, lions, tigers, buffaloes, frogs, oxen, pigs, and many kinds of birds. He incubated chunks of meat from the putrefying animal bodies in large glass flasks, while he allowed a number of large fish to decompose in bottles filled with unclean, stinking water; the odor in Dr. Redi's laboratory must have defied any attempt at description. The same kind of maggots were seen to develop in the dead tissues of many different animals, from lions to swordfish. Furthermore, several different kinds of maggots and larvae could be collected from the putrid body of a snake. This disproved the old dogma that the nature of the putrid matter determined what kinds of creatures could be spontaneously generated from it. Francesco Redi also noticed that many large bluebottle flies were buzzing about over his experimental material, attracted by its odors. He made another experiment, incubating some lumps of meat in open vessels, and others in vessels covered by a net; the result was that no larvae developed in the flasks covered by a fine-meshed net that kept the flies out. Similarly, dead fishes could be put in sealed vessels without any formation of maggots. Redi at once realized the importance of this observation: the adult flies laid their eggs on the putrid pieces of meat, and their larvae developed inside them. He also repeated some of the spontaneous generation experiments suggested by Athanasius Kircher, but he was unable to reproduce the marvelous results claimed by the bold German Jesuit.

In 1668, Francesco Redi published the book *Esperienze Intorno alla Generazione Degli Insetti,* which was published in a Latin translation 3 years later. It became one of the scientific classics, and its author was celebrated as one of the great naturalists of his time. His experiment with the bluebottle flies and the putrid meat was simple and straightforward enough to be repeated at many universities around Europe; one after the other, the biologists had to agree that he was right. Further experimental support came from the physiologist Marcello Malpighi, who had proved that the supposed "equivocal generation" of insects from the leaves of plants could be explained from the fact that the insects developed from the extremely small eggs deposited there by adult insects. The spontaneous generation dogma was also severely undermined by the growth of microscopy. Already in Jan Swammerdam's *Historia Insectorum Generalis,* published in 1669, it was pointed out that even the most minute insect, which had previously been considered little more than a speck of sand with wings, proved to be quite complicated when seen through the lens of a microscope. Like many of his contemporary biologists, Swammerdam turned his back on the absurd Cartesian view of a mechanized nature, populated with animals that were little more than automata. In particular, he viewed spontaneous generation as a godless opinion, supported only by a heathen philosophy.

The great Antonie van Leeuwenhoek was an opponent of spontaneous generation already in the 1670s, although mainly from arguments that demonstrated the inconsistency of this concept: if lice and fleas could develop from putrefaction, then why not pigs and oxen? Similarly, if frogs were spontaneously generated from mud and slime, it was inconceivable that they could all have the same structure, since the debris found in the muddy ditches was unlikely to contain all the material needed to produce a complete frog. Leeuwenhoek later became aware of the experimental work of Redi and Swammerdam, and his own microscopic investigations provided him with strong arguments against the old fallacy. In the early 1680s, having reproduced Redi's experiment with the incubation of meat in open or sealed glass tubes, Leeuwenhoek declared himself convinced that no living animal could be developed from putrefaction or from the sap and leaves of plants. Leeuwenhoek systematically studied the development of various insects, particularly fleas, flies, and aphids. He kept several fleas enclosed until he had clearly observed them lay eggs; he then carried these eggs in his pocket until they hatched. To his surprise, an intermediate stage appeared: the flea larvae, which later spun cocoons and pupated. Inside these pupas, the adult fleas could be observed. Leeuwenhoek carried several families of lice in his stockings, allowing them to feed from his own flesh while he studied their development. His wife was recruited to carry insect eggs in her bosom to keep them warm.

The discovery of the life cycles of these insects was not only of great service to entomology but also severely undermined the concept that these beings could be spontaneously generated from sweat and filth. Antonie van Leeuwenhoek described the genitals of many species of insects. He was also well aware of the existence of the spermatozoa in the semen, and their function. He was able to describe the spermatozoa of several insects, including the grain weevil and the flea. Since Leeuwenhoek believed that the size of the spermatozoa corresponded to that of the animal in question, he was astounded by the huge size of the spermatozoa of the flea. A particularly difficult matter for Leeuwenhoek was the alleged spontaneous generation of eels. Aristotle had already observed that eels had no apparent sign of spawn, milt, or sex, and up to Leeuwenhoek's time, no one had refuted him. It was not until 1691 that Leeuwenhoek discovered the uterus of the eel and demonstrated the unborn little eels therein.

When the imaginative Father Filippo Buonanni, whose theories about oceans of floating "scrambled eggs" developing into barnacle geese we have already encountered, tried to challenge Redi's experiments, Leeuwenhoek answered him in no uncertain terms. He was particularly outraged by the fact that many naturalists still differed between "complete" and "incomplete" animals. In his *Arcana Naturae Detecta,* published in 1696, after describing the structure of the fly's facet eye, Leeuwenhoek declared himself amazed that there were still people believing Aristotle's old fallacy: such a beautiful, complex little animal as a fly could never be spontaneously generated from putrefying matter, however small it might appear from the arbitrary scale of man. The British physician and poet Richard Blackmore, an opponent of spontaneous generation, was much impressed by the recent advances in microscopic science:

But the late inquirers in their glasses find
That every insect, of each different kind,
In its own egg, cheered by the solar rays,
Organs involv'd and latent life displays;
This truth, discover'd by sagacious art,
Does all Lucretian arrogance subvert.
Proud wits, your frenzy own, and, overcome
By Reason's force, be now for ever dumb.

Even Francesco Redi and his adherents had to admit certain exceptions from their anathema against spontaneous generation. Redi himself readily admitted the theory of spontaneous generation of gallflies. Many biologists were deeply puzzled by the fact that Leeuwenhoek had, with the same microscopes he had used to disprove the spontaneous generation of parasitic insects, discovered the existence of protozoa and bacteria. At the time, scientists considered that these minute beings, which could not be observed to possess any genital parts, could only be spontaneously generated. Many people were also of the opinion that lice could be spontaneously generated from human sweat and filth. In the so-called "most horrible of diseases"—phthiriasis, or the lousy disease—insects formed boils, or rather tumors, under the skin of the unhappy sufferer, whose flesh was eaten while he or she was still alive. When such a tumor burst, a multitude of crawling little insects swarmed out. The lousy disease was long considered a divine vengeance on tyrants, desecrators, and enemies of religion. Herod the Great, Sulla, King Philip II of Spain, and Pope Clement VII were considered to be among its victims. In the *Cabinet of Medical Curiosities,* I have argued that this legendary disease may well have been a reality, although it is likely to have been caused by a particular kind of mite rather than by lice.

Throughout the eighteenth century, and well into the next, there was little opposition to the concept that intestinal worms were spontaneously generated from the contents of the gut. The followers of Redi eagerly sought a way to disprove this concept, but without success; the life cycles of these worms were too complicated to be elucidated by the methodology available to eighteenth-century parasitologists. Instead, some philosophers speculated that man was in fact predestined to harbor intestinal worms, and that all species of parasitic insects and worms were meant by God to serve as his constant companions. According to these pleasant theories, already poor Adam had harbored all known species of lice, fleas, mites, and bed bugs. In addition, his intestines were full of all kinds of parasites—ascarids, tapeworm, and thread-worm, to mention but a few—all crawling merrily about in his putrid, corrupting feces. The conservative theologians were aghast at this horrid caricature of the First Man as a lousy, worm-infested vagabond who reeled about in the Garden of Eden, perpetually scratching himself, pursued by a swarm of parasites. The Italian naturalist Vallisnieri tried a compromise: in the Garden of Eden, even the crab lice and the tapeworms had been benign, friendly, and inoffensive companions who had lived together with Adam in perfect harmony. After the fall of Man, this strange truce was ended, and all the parasites were instructed by God to torment him and his descendants in perpetuity, in the most cruel fashion. Spontaneous generation was an infected topic, however, and Vallisnieri was attacked from both sides. The adamant theologians considered his musings to be sheer blasphemy, and the naturalists found his ideas not only dotty and eccentric but plainly wrong from a scientific point of view. Vallisnieri could not explain why only some individuals, and not others, were afflicted with intestinal worms; surely, if these parasites had been intended as a general scourge on humanity, no one would have been spared. Nor did his ideas take into account that different kinds of parasites were prevalent in tropical countries, compared with those occurring in Europe.

In the late 1700s, several naturalists tried to develop tapeworms by feeding animals worm eggs, but the experiments were wholly without success. In *Lebende Würmer in Lebenden Menschen,* a standard work on parasitic worms published in 1819, the German helminthologist Dr. Bremser considered the development of tapeworms the strongest evidence that spontaneous generation was a reality. It was not until the 1850s that this long-lived error in parasitology, which had done much to impede the scientific investigation of these parasitic worms, was finally disproved. By this time, several parasitologists suspected that the tapeworms were developed from a species of cystic worms named *Cysticercus cellulosae.* In the life cycle of this species, after the pig has consumed the tapeworm eggs that occur in human excrements, these cystic worms, which can be considered the larvae of the tapeworm, develop in porcine muscular tissue. When the pig is slaughtered and its pork eaten, the larvae develop into adult tapeworms inside the human alimentary canal.

The German parasitologist Friedrich Küchenmeister planned a clever experiment to test this hypothesis. Unwilling try it on himself or on some hapless medical student, he obtained the consent of the government to use a convicted murderer as a "guinea-pig." The unsuspecting murderer was given a meal of worm-infested bacon, and after he had been executed, an autopsy was performed and the intestines were carefully examined. To his delight, Dr. Küchenmeister found several young tapeworms. His great discovery was published in scientific journals all over the world, to the chagrin of another, more heroic German parasitologist, Dr. Alois Humbert, who had himself eaten *Cysticercus*-infested pork, and afterwards meticulously studied his excrements for the occurrence of young tapeworms. Dr. Küchenmeister had beaten him to the discovery of the life cycle of tapeworms by a narrow margin. In 1859, the appositely named doctor treated another German criminal to a meal of worm-ridden pork. The result was the same: a considerable number of tapeworms found at autopsy. The German government commended Friedrich Küchenmeister for his important work in solving a major health problem, but several British and French periodicals were decidedly unimpressed by his drastic method of experimentation. It was considered characteristic of the barbaric mentality of the Teutonic race that prisoners were used as "guinea-pigs" in dangerous experiments without having given their consent. In the *British Medical Journal,* Dr. Küchenmeister was harshly criticized in a leading article.

At about the time that Dr. Küchenmeister was active in the German prison kitchen, another stronghold of spontaneous generation fell in France. At the same time as the microscopic observations of Antonie van Leeuwenhoek had helped overturn the old-fashioned belief in spontaneous generation of larger animals, he was also the first person to see and describe bacteria, using his extremely powerful single lens microscopes. Throughout the eighteenth century, there had been a vigorous debate about the generation of bacteria and protozoa. The English Catholic priest John Turberville Needham, who was also a distinguished amateur naturalist and a fellow of the Royal Society, had made experiments by heating various substances, and later inspecting them closely through the microscope for the appearance of living beings. Even a vial of mutton gravy, taken hot from the fire, was observed to swarm with life after some days of incubation. Needham was opposed by another Catholic priest, the Abbé Lazzaro Spallanzani. Spallanzani was a more careful and ingenious experimentalist than Needham, and doubted the latter's claim that the boiling of a certain broth was bound to make it sterile. In spite of the experiments of Redi and Spallanzani, the early–nineteenth century biologists were still unable to explain the appearance of bacteria and infusoria in sterile broths; many of them still inclined toward the hypothesis that these minute beings, which had no discernible genitals, could develop only through spontaneous generation.

In mid-nineteenth-century France, the spontaneous generation debate was becoming increasingly heated. Madame Royer's French translation of Charles Darwin's *Origin of Species,* had been prefaced by a violent, anti-Catholic diatribe of her own concoction, and many religious people considered both Darwinism and the belief in spontaneous generation to be a serious threat against religion. In 1860, the French Academy of Sciences offered a prize for experiments that could throw new light on this age-old problem and definitely prove or disprove the spontaneous generation of bacteria and infusoria. Four years later, this prize was claimed by the great Louis Pasteur, who had performed a series of world-famous experiments. He incubated his media in flasks with long, curving necks that protected their contents from contamination with bacteria, fungal spores and other microbes. If the contents of the flask were boiled, and the neck kept intact, no microbial life appeared: but as soon as the neck was broken off, permitting contamination from the outside, microorganisms appeared. Louis Pasteur's foremost adversary was Dr. Felix Pouchet, who had proposed some kind of "vital force" in the atmosphere. He had prepared boiled flasks of fermentable matter. Of these, all contained no organisms as long as they were unopened, but once they had been opened, even on the edge of a glacier at an elevation of 6,000 feet, each developed bacterial growth. The reason for this is probably that his medium contained heat-resistant bacterial spores, while Pasteur was fortunate that his medium did not.

Louis Pasteur was widely applauded by his contemporaries, particularly his religious countrymen. At a brilliant scientific soirée at the Sorbonne, held before the social elite of Paris, Pasteur solemnly proclaimed that "Life is a germ, and a germ is life" and boasted that "Never will the doctrine of spontaneous generation recover from the mortal blow of this simple experiment." Although both Pouchet and the British biologist Henry Charlton Bastian declared themselves unconvinced by his reasoning, and defended spontaneous generation well into the end of the century, the vast majority of scientists and educated laymen supported Pasteur in the controversy. In particular, the discovery of heat-resistant bacterial endospores, by Ferdinand Cohn and John Tyndall, added further evidence to his cause, and provided a plausible explanation for Pouchet and Bastian's results.

In the sixteenth and seventeenth centuries, the philosophers and zoologists had no difficulty in classifying living things into two kingdoms: those that moved, ate things, and grew to a certain predetermined size (animals), and those that did not move or eat, and grew indefinitely (plants). The vegetable lamb and the barnacle geese were a conundrum to the early biologists, since these monstrous beings were part animals and part plants. In the mid-1600s, these creatures were fitted into a separate kingdom called zoophytes, together with certain submarine sponges. At least some of the contemporary biologists seem to have considered it quite natural that these bizarre "missing links" existed, for they provided a kind of bridge between plants and animals. At the time of Carl Linnaeus and the other great taxonomists of the eighteenth century, the vegetable lamb and the barnacle geese had already been ousted from scientific zoology, and the border between animals and plants seemed more impenetrable than ever. Throughout the nineteenth century, the various newly discovered microorganisms were distributed between the animal and the vegetable kingdoms according to the same time-honored criteria: the protozoa, which could be observed to eat and move about, were considered to be animals, while the fungi, algae, and bacteria were considered to be plants. The twentieth-century discovery of fundamental differences between procaryotic and eucaryotic cells has given the procaryotes—bacteria and some blue-green algae—a kingdom of their own called Monera. Between the higher animals on one side and the plants and fungi on the other is the kingdom Protista, containing protozoa and protophytes; within modern taxonomy, the bacteria and the unicellular eucaryotes have thus usurped the position once held by the vegetable lamb and the barnacle geese. Another interesting fact is that the old criterion that plants must be immobile is not always upheld in modern biology, as certain forms of plants actually have mobile gametes. Unlike the barnacle geese, however, these gametes have to rely on their feeble cilia and flagella for a short and jerky ride toward their dreamt-of target, rather than to fly underneath the firmament on powerful wings.

The boundary between "living" and "nonliving" has also become increasingly blurred, owing to the immense advances in molecular genetics and microbiology. The existence of viruses was first proposed in the latter half of the nineteenth century, but it was not until the advent of electron microscopy in the 1930s that these subcellular agents could be studied more closely. A virus can be defined as a core of nucleic acid (either RNA or DNA) surrounded by a protein coat. It must use the metabolic machinery of a living host organism to replicate and produce more viral particles. The French virologist and Nobel Laureate André Lwoff proclaimed that "whether or not viruses should be regarded as organisms is a matter of taste." But if they are not living, what are they? Clearly not just pieces of DNA or RNA with lumps of protein hanging about them. It is amazing to contemplate that when a virus infects a cell, its nucleocapsid is taken into the cytoplasm and the DNA or RNA is finally incorporated into the genetic material of the living cell, which is tricked into producing a multitude of viral proteins. Some of these proteins shut off the normal activity of the cell, while others are the components of the viral capsid and structure. At one stage, the "virus" is thus a stretch of genetic material, a set of proteins, and an ability to form a complete virion. It would be incorrect to claim that the virus particle is spontaneously generated, however, since its assembly is nonrandom and highly specific; viruses have millennia of biological experience.

One step further down the line are the viroids, a group of plant pathogens that challenges the concept of an infectious nucleic acid. Viroids consist only of circular strands of RNA, 250 to 575 bases long. They have no structural components, do not code for any proteins, and are dependent on the machinery of a living cell for replication. Their primitive design does not prevent these viroids from being significant pathogens: they cause the so-called "Cadang-Cadang" disease of coconut palms and the "planta macho" disease of tomatoes. The newly discovered prions at present represent the lowest form of self-replicating, parasitic life. The prion diseases are a group of neurodegenerative maladies that are transmissible through inoculation and, apparently, through dietary exposure to infected tissues. An abnormal isoform of a host glycoprotein seems to be the central, and probably the sole, infectious agent. The concept of a protein as the sole component of the transmissible agent of disease would have seemed ludicrous just 10 years ago but is today taken quite seriously. For many years, a classic example of prion diseases was Kuru, a syndrome of slow degeneration of the brain transmitted by the practice, in parts of New Guinea, of eating the brains of deceased relatives in certain rituals. Today, by far the most familiar of these diseases is bovine spongiform encephalopathy, or mad cow disease, which put the British nation in a state of turmoil. This disease was caused by the feeding of cattle with dietary protein supplements that had been "enriched" with offal from sheep afflicted with scrapie, another prion disease. The feeding of unsuspecting cows with ground carcasses from other herbivores would seem nearly as unnatural as eating raw, uncooked human brains; indeed, the ultimate effect has been exactly the same, since the prion protein is transmissible by the oral route. Several people have been affected with a new kind of encephalopathy, distinct from the older, known forms of spongiform encephalopathy, but with several clinical and laboratory characteristics resembling bovine spongiform encephalopathy transmitted to a cat or macaque.

Vital Dust

Today, the old zoophytes are completely forgotten, except by lovers of curiosities and professional historians of zoology. The gluttonous Franciscan friar who gorged himself with delicious roasted goose in front of his fasting brothers, claiming that he was really eating shellfish and vegetables, would probably be forcibly evicted from the monastery, with the goose thrust up under his gown. Travelers roaming the southern provinces of the former Soviet empire in search of a plate of tender lamb cutlets would be well advised to travel equipped with an internationally valid credit card, rather than to brandish a spade and a chainsaw. Although its most extravagant offspring—the vegetable lamb and the barnacle geese—are now past recollection, the original concept of spontaneous generation, as formulated by Empedocles and Aristotle more than 2,400 years ago, is still very much alive, although in a different guise.

During the French debate about spontaneous generation, the strongest argument proposed by Pasteur's adversaries was that if living beings could not be generated from nonliving matter, how had life on earth originated in the first place? This question is still valid for those who do not accept the dogma of divine creation. An alternative hypothesis, first proposed by the German H. Richter in 1865 and later, in more detail, by the Swedish chemist Svante Arrhenius, was that life had originated elsewhere in the galaxy and had been transmitted to Earth by space-borne microorganisms. Since the time of Arrhenius, several different interpretations of this so-called *panspermia hypothesis* have been put forth. *Lithopanspermia* is the distribution of microbes or their spores by meteorites and *radiopanspermia* the dissemination of living organisms by stellar radiation. The most fanciful variant, *directed panspermia,* proposes the deliberate seeding of life on Earth by a spacecraft manned by highly developed extraterrestrials. All these variants see life on Earth as one small part of an ubiquitous "galactic life." The acceptance of the panspermia hypothesis would depend on the discovery of live microorganisms in interstellar space, something which has not yet been achieved. Furthermore, many astrophysicists doubt that any living creature could withstand the cold, radiation, and vacuum effect in outer space. Nor does the panspermia hypothesis address the question of where—and how—this galactic life first developed.

In the original 1859 edition of *The Origin of Species,* Charles Darwin avoided discussing spontaneous generation and the origin of life, apart from suggesting that all life on Earth evolved from a common ancestor. Cautious not to offend mainstream religious opinion more than he already had, he wrote that "the Creator" had originally breathed life "into a few forms or into one." He did not, at this time, envisage that the origin of life could be rationally analyzed. The German physiologist and philosopher Ernst Haeckel was a supporter of many of Darwin's ideas. He had rather a speculative turn of mind and was a champion of monism, the unity of mind and matter. In a series of lectures on Darwinism held during the winter semester of 1865 to 1866, Haeckel did not hesitate to suggest that the first living organism on Earth had formed through spontaneous generation—first by a gradual synthesis of increasingly complex molecules, later by the formation of a *moner,* or complex plasma clump, which Haeckel considered a forerunner of the primitive protocell. Some inspired speculations by Darwin about the origin of life may well have been influenced by Ernst Haeckel's writings. In 1871, he wrote in a private letter that if "some warm little pond, with all sorts of ammonia and phosphoric salts, light, heat, electricity, &c., present" existed, any protein compound formed would be likely to undergo further metabolism rather than to be degraded, since there was at this time no other living matter. Darwin's eloquent grandfather, Dr. Erasmus Darwin, had prophetically expressed a similar opinion in his *Temple of Nature:*

Then, whilst the sea at their coeval birth,
Surge over surge, involved the shoreless earth;
Nursed by warm sun-beams in primeval caves
Organic life began beneath the waves,
Hence without parent by spontaneous birth
Rise the first speck of animated earth.

In 1924, the Russian biochemist Aleksandr Ivanovich Oparin formulated an alternative hypothesis. He was a fanatical communist, and took part in the Bolshevik publishing program, which poured out a steady stream of materialist, antireligious pamphlets about Darwinism and human evolution, for the edification of Russian laboring men. At 30 years of age, Oparin addressed the question of the origin of life in a booklet published by the *Moscow Worker*. In this early work, Oparin proposed that life had evolved through the spontaneous generation of colloidal gels that had the basic properties of life. Life on earth had appeared suddenly, by chance, but in a way completely explicable in terms of physics and chemistry. Even a firebrand like Oparin had a certain grudging respect for Louis Pasteur, and he had to concede that the experiments of this great Frenchman had conclusively proved that spontaneous generation was no longer a reality. Instead, Oparin cleverly proposed that once life had appeared on earth, further spontaneous generation was precluded, since the colloid precursors in question would be consumed by the life that already existed. He did not quote the theories of Ernst Haeckel, of whose work he had probably never heard.

Oparin published further articles on the same theme in serials such as *Herald of the Communist Academy* and *Under the Banner of Marxism.* Since Oparin's treatises were quite advanced, he apparently thought highly of the intellectual ability of the average Russian worker; it is astounding, in fact, that his pamphlets were quite popular. This was a time when there was still widespread, true enthusiasm for the sake of communism and the ideal new state that was emerging from the remnants of the crumbling empire of the czars. One suspects, however, that the lack of other interesting reading material in that country also played a part. Works on natural history probably assumed a new luster when compared to reports of the endless speeches at the Komsomol's annual jumble sale or the Communist Women's League's march to celebrate the revolution.

Aleksandr Oparin was frequently distracted by the insistence of his superiors that he should devote his energies to more utilitarian fields of research. Although he could not afford to openly disobey his masters — one of his pamphlets dealt with the biochemical aspects of tea production—his mind was set on trying to elucidate the origin of life. In a later version of his book, published in 1936, it is apparent that Oparin had thought the matter through more closely. He now attacked the concept of spontaneous generation and saw life on earth as a natural stage in the evolution of matter. The first living thing on Earth, which could itself procreate its species, had developed as the result of a long, time-consuming series of chemical reactions, resulting in the formation of increasingly complex molecules. Oparin was aware of the existence of viruses and bacteriophages, and he proposed that a long evolutionary process had led to increasingly complex agglomerates of molecules, finally resulting in the formation of viruses and bacteria. His early supporter, J. B. S. Haldane, agreed with the early virologists, claiming that these noncellular forms of life could be spontaneously generated. He proposed that life had then remained in the virus stage for many millions of years, before a random assembly of elementary units resulted in the formation of the earliest protocell. Oparin's book was translated into English in 1938, but in spite of Haldane's support, it took considerable time for these novel ideas to be taken seriously by the Western scientists. Oparin was considered a dangerous, harebrained outsider, and it did not aid his cause that his book was obscurely written and overflowing with ideological language. Many people refused to believe that any worthwhile scientific research could emerge from communist Russia, and it was to take several decades until anyone attempted to verify Oparin's hypothesis experimentally.

Throughout his long life, Oparin remained a righteous communist; if he had been with us today, he would probably have been surprised to hear that his theories about the origin of life have proved more tenacious than the political system he was supporting. He was even a henchman of the notorious Lysenko, who denied the existence of genes and described genetics as a capitalist, bourgeois enterprise linked to fascism. Oparin fully shared Lysenko's distrust of modern molecular biology and genetics. He was frequently led astray by political concerns, and Lysenko made use of him in various unsavory intrigues. In 1950, Oparin was pressed by Lysenko into awarding the Stalin prize to a certain Olga Lepeshinskaya for her work on the spontaneous generation of cells from nutrient media. Both before and after, Oparin strongly denied the spontaneous generation of cells, and the Stalin prizewinner showed him no gratitude, later attacking him with impunity. In spite of Oparin's steadily growing prestige internationally, the rigid Communist system had taken its toll on his creative powers, and much of his time was taken up by administrative affairs and endless squabbles with envious, intriguing colleagues. Although he lived until the year 1980, Oparin did little worthwhile research during the last 40 years of his life.

Experimental studies of molecular evolution have established that the amino acid sequences of macromolecules in various organisms are much closer related than could be expected merely from chance: the same protein can be found in a corn plant, a microbe, a bullfrog, and a human being. This provides strong evidence for the unity of life—that all living organisms on Earth evolve from one common ancestor. It is not known exactly when this earliest living being came into existence. The planet itself was condensed from a huge cloud of gas about 4.5 billion years ago. The earliest known microfossils, from Warrawoona in Western Australia, are 3.5 billion years old. These fossils contain bacterium-like bodies and laminated sedimentary structures resembling algal mats. Some of them resemble extant chemoautotrophic and photoautotrophic procaryotic morphotypes. Studies of carbon isotype rations of kerogen in the Warrawoona sediments have indicated that photosynthesizing, or even cyanobacterium-like, organisms were active as early as 3.5 billion years ago. These bacteria are quite highly advanced and are capable of oxygen-producing photoautotrophy. This would imply that life on earth had, at this stage, already existed for a prolonged period of time, as would the striking morphological diversity of the species identified in the Warrawoona sediments.

A recent physical analysis of certain western Greenland fossil carbon deposits older than 3.85 billion years provide evidence that life on earth existed already at this time, since there is a predominance of the lighter over the heavier carbon isotype, a characteristic feature of biological carbon assimilation. No known process other than life on earth would explain these data, unless there was at this time some obscure abiotic process that produced isotopically light carbon and selectively incorporated it into apatite grains, which would seem unlikely. The advanced state of the 3.5 billion-year-old Australian microfossils agrees well with the concept of primitive life existing 3.85 billion years ago. Since the planet is assumed to have been quite uninhabitable during its first 500 million years of existence, owing to the continuous bombardment of asteroid showers and violent volcanic eruptions, it may be hypothesized that the common ancestor of life on Earth appeared between 4.0 and 3.85 billion years ago. An intriguing question is how pervasive primitive life was in our solar system at about this time. A Martian meteorite in the Antarctis has been demonstrated to have possible fossil remains of post-Martian biota. Even more intriguing, a recent analysis of the asymmetry of the extraterrestrial amino acid enantiomers on the so-called Murchison meteorite, which fell in Australia in 1969, shows an enrichment of left-rotational isoforms, just like the amino acids that are components of life on Earth. This suggests the hypothesis that some process favoring left-rotational isoforms was in operation before the appearance of life on earth, and possibly before the formation of our solar system.

The mechanisms of the emergence of life on earth still remain obscure. The earliest common ancestor of life had to possess genetic information for functioning and reproduction as well as the means to replicate and carry out these instructions. Its system of replicating its genetic material had to allow for some random variation, allowing the evolution of new traits, and, with time, the development of more accomplished forms of life. This ancestor of life had to be nonparasitic, since there was nothing else living, which rules out viruses and viroids playing any role, at least in the forms they have today. Both Aleksandr Oparin and J. B. S. Haldane were well aware that, if the prebiotic atmosphere had been as oxygen rich as today, there was no possibility for the organic compounds needed for life being formed on earth. They instead proposed that the atmosphere on the young earth was a reducing one like that of the outer planets, containing very little oxygen and instead being rich in hydrogen, methane gas, and ammonia. The earliest attempt to test Oparin's hypothesis was made in 1953 by Stanley L. Miller, a graduate student in the laboratory of the Nobel Laureate Professor Harold C. Urey. Miller constructed an artificial "atmosphere" of methane gas, ammonia, water vapor and hydrogen gas above an "ocean" of water. He then subjected the gases to repeated electrical discharges to simulate lightning. The result was that not less than 10% of the carbon in the system became a constituent of various organic compounds; even more remarkably, many of the twenty standard amino acids were formed.

This experiment has since been reproduced many times, with varying compositions of gases. Although there has been some opposition as to the true nature of the prebiotic atmosphere, mainly that it probably contained less hydrogen than presumed by Miller and Urey, the validity of their findings has not been seriously challenged by later investigators. Indeed, another analysis of the Murchison meteorite has shown a spectrum of amino acids that is remarkably similar to that analyzed by Miller, both in nature and in relative quantity. Thus, amino acids are among the most conspicuous products of abiotic chemistry, both on Earth and in space. It has also been shown that these amino acids link together on heating of the primeval so-called "hot dilute soup," although generally not in a way reminiscent of today's proteins or oligopeptides. In the experiments of Professor Sidney Fox, the heating of a solution of amino acids to 170°C resulted in the formation of so-called proteinoid, micellar structures with a certain catalytic ability. Short peptides have also been shown to form through thioester bonds under conditions less extreme than in the Fox experiments. These primitive proteins are nothing like the highly specialized, present-day enzymes and catalysts, however. It is extremely difficult to envision a system in which proteins could replicate in the absence of nucleic acids.

Indeed, there is much to be said for the concept that proteins must have been preceded by ribonucleic acid (RNA). In biochemistry as it is today, proteins are always coded for by RNA. An RNA molecule can be described as a string of ribose (sugar) and phosphate groups, to which are linked four different kinds of purine or pyrimidine bases: adenine, guanine, cytosine, and uracil. These four bases constitute the alphabet of the genetic information, and triplets of them form the words: for example, the triplet cytosine-uracil-uracil instructs the protein-forming ribosome in the cell to add the amino acid leucine to the growing strand of protein. Several distinguished biologists have proposed that an "RNA world" of gradually more and more accomplished, self-replicating RNA molecules preceded the first nucleic acid–encoded protein synthesis. Synthesis of increasingly complex protein enzymes then gradually evolved, until the age of the protocell, in which a functional system of RNA and protein was enveloped by a (probably lipid) membrane, was reached.

One major difficulty of this concept of an RNA world as the earliest self-replicating form of life is that, today, protein enzymes are needed to form complete RNA molecules. According to Crick's central dogma, information always flows from nucleic acid to protein, never in the reverse direction; hence, RNA came before protein. The concept of an RNA world received important support in 1983, when the American scientists Thomas Cech and Sidney Altman discovered that certain RNA molecules, so-called *ribozymes*, were endowed with catalytic ability. RNA enzymes such as these could thus carry out tasks in RNA processing that are today performed by proteins. Indeed, experiments by Thomas Cech and Jack Szostak have demonstrated that certain modified ribozymes can carry out important steps in RNA processing, such as the splicing together of strings of oligonucleotides. Although no RNA-catalyzed reproduction of RNA has yet been demonstrated, it does not take much imagination, given these important results in ribozyme biochemistry, to envision a self-sufficient RNA world, in which catalytic proteins played a subordinate part.

A second, more serious argument against the concept of the RNA world is that the biosynthesis of oligonucleotides is a quite complex process. It is true that the American biochemist Juan Oró demonstrated that adenine, one of the four important bases in the RNA molecule, could be formed from an aqueous solution of hydrogen cyanide and ammonia, and that other researchers have derived the other three essential RNA bases in a similar manner. It is also possible to produce ribose from formaldehyde. All these reactions have low yields, however, particularly in prebiotic conditions. Nor has there been any wholly convincing model for the assembly of nucleotides from ribose, phosphate, adenine, or other bases, least of all of the formation of long (50–60 residues) polymers of these nucleotides. A formation of such a complex molecule could not occur through mere chance or spontaneous generation, since it would not be enough to have just one or two residues formed: the same, very unlikely combination of molecules would have to occur time after time in an orderly manner. One daring hypothesis to overcome this difficulty is that RNA was preceded by some other, similar molecule carrying genetic information. It has even been suggested that a peptide string with attached RNA bases was once the transducer of the genetic code. But there is no trace of any molecule like that even in the most primitive organisms of the present time, which would have been expected if it once had played an important role in the genetical signal transduction of living cells.

In *Vital Dust,* his brilliant review of life on earth and its origins, the Nobel Laureate Professor Christian de Duve advanced a novel theory. The conditions described by Miller could well allow the formation of catalytic multimers of hydroxyacids and amino acids. These multimers might, under certain conditions, perform many of the tasks of present-day enzymes and catalyze the interactions between different varieties of RNA and protein. Another theory was put forth in 1996 by Professor Leslie E. Orgel and coworkers. Certain minerals and clays can function as catalysts and facilitate the polymerization of amino acids and nucleotides, by binding and supporting oligomers. Whereas in an aqueous solution, as most decamers are formed, the presence of mineral (hydroxylapatite or illite) in the case of amino acids, or clay (montmorillonite) in the case of nucleotides, allows the formation of polymers of 50 to 55 residues. A polynucleotide of this size can function as a small ribozyme, and these ingenious experiments have considerably strengthened the concept of a self-sufficient RNA world.

A Summing Up

When I first read a major textbook on the history of science, as a medical student 14 years ago, the story of spontaneous generation seemed wholly ludicrous, as it would indeed appear to anyone who surveys only its most extravagant offspring. What has been less appreciated is the major impact this doctrine has had on sixteenth- and seventeenth-century biology. Up to about the year 1670, the almost universal acceptance of spontaneous generation is striking, as is the number of books and theses written on this subject. The verdict of the traditional, positivist historiography of science is that the great Francesco Redi first discredited spontaneous generation by his clever experiments, and then the great Louis Pasteur finally refuted this laughable error, which had originated with the wicked, unreliable old Aristotle. The true story, which does not readily allow itself to be condensed into a few banal sentences, is of course much more complicated than that. It is indubitable that Redi deserves much credit for his experiments, but they only aimed to discredit the spontaneous generation of certain insects; indeed, Redi was himself a supporter of the spontaneous generation of gallflies and intestinal worms. At the time of the early microscopists, such as Swammerdam and Leeuwenhoek, late-seventeenth-century science had definitely caught up with some of the most absurd and easily refutable elements of the old dogma. The vegetable lamb was also an early casualty of late-seventeenth-century scientific zoology.

Throughout the eighteenth and nineteenth centuries, the spontaneous generation controversy was vigorously debated. The microscopic observations of Leeuwenhoek and others had opened another microcosm—the world of protozoa and bacteria—and until the late nineteenth century, biological research could not provide an explanation of the means of reproduction of these microscopic beings. More astonishingly, some of the traditional tenets of spontaneous generation were also remarkably long-lived: the equivocal generation of intestinal worms and parasitic insects was debated well into the mid-1800s, and the barnacle geese were not permanently laid to rest until the 1780s. Even after Louis Pasteur had, with a mixture of skill and luck, been able to discredit the spontaneous generation of bacteria, the old concept died hard. Henry Charlton Bastian, for example, kept up a steady barrage of books and articles in its favor well into the twentieth century. Today, no one but a crank would advocate the spontaneous generation of protozoa, bacteria, or viruses; the advances of modern biology have made their methods of replication well known. The blurring of the concept of "living" and the discovery of novel forms of parasitic 'life' has again raised the specter of spontaneous generation, but no such process has as yet been proved to occur even in these extreme instances.

Spontaneous generation, in the sense that the first living being, the ancestor of all life, was formed from nonliving matter, is very likely to have occurred about 3.85 to 4.0 billion years ago. It was not a strictly *spontaneous* or random combination of molecules, however, but a slow, painstaking, gradual formation of increasingly complex molecules.

Odd Showers

❧ *The land frogs are some of them observed . . . to breed by laying eggs, and others to breed of the slime and dust of the earth, and that in winter they turn to slime again, and that the next Summer that very slime returns to be a living creature; this is the opinion of* Pliny; *and* Cardanas *undertakes to give reason for the raining of* Frogs; *but if it were in my power, it should rain none but water* Frogs, *for those I think are not venemous . . .*

Izaak Walton, *The Compleat Angler* (1653)

IT IS TOLD IN THE HOLY SCRIPTURE THAT NOT LONG after the Children of Israel had made their escape from the pursuing host of the pharaoh, they reached the desert of Sur. After wandering about in this desert for 3 days, the dehydrated Israelis wanted to drink water from a stream, but the water proved to be bitter and unfit for human consumption. The people murmured against Moses, complaining of their thirst. Moses called to the Lord, who showed him a certain kind of wood. Moses threw the wood into the water to make it sweet, and the people were finally able to quench their thirst. Later, when they had reached the desert of Sin, the

Children of Israel were tormented by hunger. The assembled people again murmured against their saviors, Moses and Aaron, with the bitter words "Would to God we had died by the hand of the Lord in the land of Egypt, when we sat by the flesh pots, and when we did eat bread to the full; for ye have brought us forth into this wilderness, to kill all this assembly with hunger!" Moses, whose patience with these ungrateful ones must be admired, again saved the day: the Lord let a rain of manna fall on the camp of the Children of Israel, and fallen quail could be collected in its vicinity.

Generations of schoolchildren have been taught to believe these marvelous happenings, no doubt imagining a shower of nourishing cereal and tender roasted quail slowly descending toward the startled Children of Israel, who ran excitedly about and collected the food in their tunics and headcloths, praising the Lord for this great miracle. The children would have been even more impressed if their schoolmaster had pointed out that fish, frogs, toads, and worms could also fall from the sky with rain. If one of the grumbling, dissatisfied Children of Israel had murmured against Moses, requesting a plate of delicious *grenouilles* to top off his dinner of quail and manna, there was every chance that even this cantankerous gourmet would receive his desired sustenance from above.

These anomalous showers of living animals or other unexpected substances from the skies have fascinated humankind since the beginning of time. They were long considered as portents: a shower of blood foreboded war and bloodshed throughout the land, while a shower of worms was a sure sign of divine displeasure. Throughout the sixteenth and seventeenth centuries, there was a vigorous debate, among some of the foremost scholars of the time, about the cause and mechanism for these rains of living animals. Many ingenious hypotheses were put forth, and showers of fish, frogs, and toads were described in the some of the most prestigious scholarly journals of the time. As late as the mid-1800s, the existence, or nonexistence, of these odd showers was vigorously debated in various periodicals. At the end of that century, however, the vast majority of scientists had declared these occurrences manifestly impossible and deplored that the vast majority of the ignorant populace still supported this vulgar error. This churlish attitude toward anomalous precipitation has remained well into the twentieth century: instead of being discussed in the *Philosophical Transactions* of the Royal Society of London, or the *Mémoires* of the French Academy of Sciences, however, the showers of fish or toads occurring in the 1990s are derided in the "funny pages" of the tabloid newspapers.

Strange Rains in Antiquity

In 1555, Olaus Magnus, the last Catholic Primate of Sweden, published his *Historia Gentibus Septentrionalibus,* a monumental chronicle of the ethnology, geography, and cultural history of Scandinavia. After the Reformation, Olaus Magnus was expelled from his cathedral in Stockholm and dedicated the remainder of his life to theological and scholarly pursuits. His chronicle was thorough and well written, with many quaint details; it deservedly became quite popular and gave rise to a considerable interest in Scandinavia among European scholars, who had previously cared little for the state of these cold, desolate isles of the North. An entire chapter of the *Historia* is devoted to the rains of fish, frogs, rats, worms, and stones that were of frequent occurrence in the Nordic countries. Olaus Magnus believed these odd showers to be caused by the inner warmth and stickiness of the Scandinavian clouds, which led to the spontaneous generation of many different substances, including living animals. According to an old Swedish chronicle, blood rained down at Ringstadholm in 1316, a prodigy considered by some to portent the civil war between King Birger and his brothers. In 1473, a deluge of worms rained down from the sky in Westgothia. These worms devoured the crops in the fields, and the people considered them a punishment sent by God. A clever peasant named Tordo Jeppson advised that solemn prayers should be said to St. Catherine of Wadstena, a local patron saint. After a cow had been sacrificed as a votive gift, the peasant succeeded in exterminating the worms; this odd happening was considered one of St. Catherine's foremost miracles. According to the Swedish chronicle of Joen Petri Klint, blood rained down in the town of Söderköping in 1562, 1565, and 1567; in the latter year, this town was raided by the Danes and its castle burnt to the ground.

In his *Natural History,* Pliny detailed the Roman chronicles of odd showers. When Manius Acilius and Gaius Porcius were consuls, milk and blood once rained from the sky; on several later occasions, showers of meat, wool, and bricks were observed. In the year 53 B.C., iron rained from the sky near Lucania; the year after, General Marcus Crassus and his entire army, containing a large contingent of Lucanian soldiers, was annihilated in the war against the Parthians. Shortly before the death of Emperor Nero in A.D. 68, a rain of blood fell with such force that streams of blood and gore flooded the streets of Rome.

Pliny quoted several older sources on odd showers in ancient times, few of which are available today. It seems as if the existence of showers of live animals was a subject of debate already in the fourth century B.C. In a note "on animals that suddenly become visible," Theophrastus of Eresus, the pupil and successor of Aristotle, wrote that little frogs and snails could not, as many people thought, fall with the rain. Instead, their holes in the ground were flooded with water, forcing the animals out from their lairs in large numbers and giving the illusion that they had been falling with the rain. The philosopher Claudius Aelianus is unlikely to have been convinced by these arguments, since he had himself once witnessed a brisk shower of living small frogs while traveling from Naples to Puteoli. He wrote that the foreparts of these supernatural animals crawled, supported by two feet, while the hind part ended in a tail; based on his description, they are likely to have been half-developed tadpoles.

The Greek polymath Athanaeus of Naucratis, who flourished about A.D. 200, wrote a fifteen-book chronicle called the *Deipnosophists,* or the Banquet of the Learned, in which he quoted the works of more than eight hundred earlier authors whose works he had consulted at the then extant Alexandrian Library. One of his subjects of interest was the strange rains of olden times. According to a certain Phoenias, who had written a chronicle on the Eresian Magistrates, it once rained fish continuously for 3 days in the environs of Chersonesus. Another old writer, Phylarchus, knew of several prodigious rains of fish, frogs, and wheat. The strangest rain of all struck the inhabitants of the cities Peonia and Dardania: frogs rained down from heaven for several weeks. So great was their number that the streets and houses were full of them, and the inhabitants had to shut up their doors and windows. Even this measure was to no avail, however, since the frogs gained an entry into all the houses and even crept into the kitchen utensils; every meal thus had the distinctive taste and smell of none too fresh *grenouilles.* The people were not able use any water, since it was poisoned by the rotting corpses of the frogs that had fallen into their tanks, and they could hardly put their feet on the ground, owing to the profusion of frogs. In the end, they had to leave their native lands and move to another part of the country because of the nauseating smell of the multitudes of dead frogs.

Prodigious downfalls of a similar kind often occurred in the British Isles in ancient times, at least if Thomas Short's odd *History of the Air* is to be trusted. In the year A.D. 4, "it rained blood above 5 hours in London," a phenomenon that would stir up considerable controversy and speculation were it to happen today. He also recorded that, in A.D. 89, blood rained for 3 days in England. Other notes in Short's list tell us that in 766, it "rained blood

3 Days, then venemous flies, then Mortality," and that in 1014, a heap of clouds fell and smothered thousands. A famous old Irish chronicle, the *Annals of the Four Masters,* reports that in A.D. 690, "it rained a shower of Blood in Leinster this year. Butter was there also turned into lumps of gore and blood, so that it was manifest to all in general. The wolf was heard speaking with a human voice, which was horrific to all." Twenty-six years later, in A.D. 716, Leinster was struck by another shower of blood; in the same year, silver rained in Othain-mor and honey in Othain-Beag.

Olaus Wormius, Carl Linnaeus, and the Lemmings

In his *Historia,* Archbishop Olaus Magnus described another famous Scandinavian natural curiosity: the mice, or lemmings, that rained down from the skies in Lapland. They fell in huge showers, like the Egyptian locusts, and devoured all vegetation to be found in these barren mountains. The lemmings were feared by the Lapps and their dogs, since these supernatural rodents had a punishing bite: every creature touched by their poisonous, sharp teeth soon after expired in agony. These armies of airborne lemmings ruled Lapland like an enemy host: they marched about at will, laying waste the countryside and driving the wretched Laplanders before them. At a given time, the rodents all died, or were eaten by ermines and foxes. The Laplanders exulted, happy to be free of these murine invaders from the clouds—but not for long: the noxious decaying bodies of the lem-

Lemmings fall from the sky in Sweden; they march forth in a column, but are attacked by ermines, one of which is caught in a trap. An illustration from the *Historia Gentibus Septentrionalibus* of Archbishop Olaus Magnus.

A Scandinavian shower of fish, from the *Historia Gentibus Septentrionalibus* of Olaus Magnus.

mings spread a dreadful contagion, afflicting the Lapps with "vertigo and jaundice." It has been demonstrated that Olaus Magnus fetched the major part of his chapter on these terrifying, airborne killer lemmings from another contemporary source, the German Jacob Ziegler's *Scondia.* Some historians have suspected that Olaus Magnus plagiarized the whole section from this work, without acknowledgment, and that the venerable archbishop had not even seen a lemming. This would not have been an isolated instance, since Olaus Magnus freely incorporated large chunks of other geographical works into his *Historia.* It has, in particular, been pointed out that the drawing of the lemmings in *Historia* erroneously depicts them as black with a long tail; it is likely, however, that the drawing was not made by Olaus Magnus personally, but by some individual with little zoological knowledge. In his text, the archbishop quite correctly describes the lemmings as speckled in color, and the size of a vole. In the years 1518 and 1519, Olaus Magnus had traveled widely in Lapland, while employed as the secretary of the notorious Arcimboldus, a reckless Roman seller of indulgences, whose excessive trade in absolution is said to have precipitated the Swedish Reformation. Olaus Magnus was a man of observation, and the sale of indulgences to the sinful Swedes of the Northern outback is unlikely to have entirely prevented him from studying the Laplandish fauna.

Professor Olaus Wormius, the noted anatomist and zoologist of Copenhagen, was one of the many naturalists who had Olaus Magnus's description of the remarkable habits of life of the lemmings. His Norwegian friends had supplied him with further fanciful accounts of lemmings falling from above. In 1578, a great number of large, yellow mice were said to have fallen near the city of Bergen. In 1651, the parish clerk of Ravnefjord was traveling in a small boat with some friends, when two lemmings fell from the sky into the boat. Soon after, a powerful thunderstorm burst on them. A no less remarkable observation was made by an invalid old lady, who had just left her sick bed for a nip of fresh air; she received the shock of her life when a lemming fell from a clear sky into her apron. From the Bishop of Bergen, Olaus Wormius had received a "pickled" lemming, which he dissected with great care, together with his friend, the celebrated anatomist Thomas Bartholin. To their surprise, they found that the animal's organs of generation were quite normal, arguing against that the lemmings were spontaneously generated in the skies.

In his 1653 treatise on "The Norwegian Mouse," Olaus Wormius still accepted the rains of lemmings as a fact; in the manner of his time, he valued the testimony of older authorities more than his own observations. He was undecided whether the lemmings were created de novo from decaying substances in the clouds, or if they were transported from faraway places by the wind. He rather leaned toward the latter theory, quoting earlier observations of fish and frog rains, but was at a loss to explain why only lemmings, and no other Laplandish animals, were falling from the clouds. In a section on supernatural aspects of lemmings, Olaus Wormius quoted the belief, current among the common people, that the lemmings were sent as a divine punishment to make the sinful people repent their wicked ways. The Norwegian clergy were not without means to combat the strange, alien mice, however: Olaus Wormius quoted several curses and exorcisms to be used in the churches to drive the supernatural rodents away. Sometimes, the entire congregation knelt, trembling, before the local vicar, as he pronounced a solemn malediction on the invading lemmings. Olaus Wormius gave the lemming's skeleton to the Duke of Holstein-Gottorp, who wanted it for his private museum; a figure of it was reproduced in the treatise of Olaus Wormius, accompanied by a Latin epigram by Thomas Bartholin:

> *Qui pluit ex coelo repetit vestigia coeli:*
> *Hoc opus est Wormi, quod fuit ante Iovis.*

The lemming, presumed to have fallen from the sky, has now "again sought the traces of heaven," since it was put in an exalted position in the duke's museum. Jove was associated with rain, and his "opus" in sending down the lemming was taken over by Ole Worm (Wormius), who anatomized the animal and made it fit for the ducal repository. A translation can be given as follows:

The Lemming, fallen in a storm,
Has now retraced its path up high;
The work of Jove in the far sky,
Has been usurped by Ole Worm.

While traveling in Holland, England, and France in the 1720s and 1730s, Carl Linnaeus was often asked by other naturalists whether it was really true that mice fell from the clouds in Lapland. His answer was noncommittal: in spite of his respect for the learning of Olaus Wormius, Linnaeus was unwilling to accept these wonders without having witnessed them himself. In 1733, Linnaeus made an extended tour of Lapland, one of the objectives of which was to study the behavior of the lemmings at close range. He was surprised to see that the lemmings occurred quite commonly in large parts of Lapland: their burrows could frequently be observed near grassy hillocks. The lemmings were not easily frightened: when Linnaeus teased one of them, it barked like a puppy and bit his stick. Linnaeus observed that the female lemmings often had five or six young in their dens, and that they had eight teats, just like other mice. Linnaeus studied the anatomy of several pregnant lemmings to observe the development of their young: he could thus definitely disprove the tale of their spontaneous generation in the skies.

Carl Linnaeus was, in general, an opponent of spontaneous generation, and he declared that although Olaus Wormius and other seventeenth-century writers might have supported this ancient doctrine, modern front-line scientists had to maintain a skeptical attitude. Linnaeus was thoroughly unimpressed by the Laplanders—these truculent Scandinavian aborigines—who were a mine of disinformation on various aspects of natural history. They maintained, as they had probably done to Olaus Magnus 215 years earlier, that the lemmings had a poisonous bite, and that they could not be eaten. Linnaeus observed, however, that dogs and foxes ate them with alacrity and that humans could probably do so too; he did not try the experiment himself, but merely observed that "no four-footed beast is too poisonous to be eaten, as is proven by the Chinese, who eat all kinds of animals." The Laplanders also claimed that, in spite of the venomous nature of these rodents, the reindeer liked to gobble up the lemmings as a welcome variation to their herbivorous diet. Linnaeus remained incredulous, however. The Laplanders finally tried to make the famous doctor from Stockholm believe another tall tale: that both men and reindeer could be sucked up by the thick clouds surrounding the mountain peaks and transported widely before being put down on another mountain far away. This time, Linnaeus tried the experiment, wandering off on a tall mountain. He wrote that the sensation of being wholly enveloped by the cloud far up on the mountain was an eerie one: he could hardly see his hands before him. His hopes of an early return to Stockholm as a passenger on a cloud were thwarted, however, and he was fortunate to escape this adventure without falling down a deep precipice or into some abyss covered by the light snow. After this experience, it was no wonder to him that the Lapps and their animals tended to disappear while walking these winding, perilous mountain pathways, with almost zero visibility.

Carl Linnaeus was well aware of the reason for the fable of the aeronautic "Norwegian mice"; it was the strange phenomenon of lemming migration, which occurred about once every 10 to 20 years. Thousands of lemmings migrated down the mountains in strict marching order, eating every green thing in their path. Linnaeus wrote that "if they encounter a lake, they do not make a detour around it but instead swim it, even if they have to cross at its widest diameter; if they find a boat in their way, they do not escape but instead endeavor to climb onto it, but only to leap into the sea on the opposite side." In Sweden as well as in Norway, lemming migration had been considered a portent: the Dean of Lima told Linnaeus that he could recall that, many years ago, the entire parish had gathered to say prayers against the invading, supernatural rodents.

Frogs from the Sky

In Thomas Short's *History of the Air,* it is briefly stated that in 1346, "it rained Toads and Rain mix'd." This was probably the same anomalous downfall described by the sixteenth-century German chronicler Conradus Lycosthenes in his *Prodigorum et Ostentorum Chronicon.* A quaint illustration depicts the amphibians being formed in the clouds and diving down toward the earth when fully formed, to land safely on grassy meadows. The Renaissance scholars no longer considered the rains of live animals as portents of divine displeasure; they instead endeavored to find a rational explanation for these extraordinary phenomena. The old doctrine of spontaneous generation provided one: it was easy to imagine that the toads and frogs were generated from putrid matter in thick clouds. They were hovering above, adhering to the cloud until being fit to fend for themselves in their natural habitat; the frog-clouds then released their numerous brood, which fell down with the rain. Although this theory may seem as hazardous as the wingless flight of the frogs toward the tempting ponds and marshlands on Earth, it was ac-

A shower of frogs occurring in the year 1345, as figured in the *Prodigorum ac Ostentorum Chronicon* of Conradus Lycosthenes.

cepted by several early adherents of the spontaneous generation doctrine. Along with the vegetable lamb and the barnacle geese, the falling frogs and fish were considered among the foremost evidence that animals could be formed from vegetable or putrefying matter. The Jesuit polymath Gaspar Schottus, for example, wrote in his *Physica Curiosa* that "little frogs are not only formed through the action of rain on dust, but these animals also often fall to the ground with rain, having been generated in the clouds. I know of respectable witnesses to these occurrences; indeed, I have seen it happen myself."

As the years went by, knowledge in amphibian physiology increased. It did not require a brilliant mind or great knowledge in natural history to sit down by a pond to watch frog spawn develop into tadpoles and then into adult frogs. These observations proved that amphibians usually developed in a normal fashion, and thus spoke against their regular descent from the sky fully formed. The Italian mathematician and naturalist Girolamo Cardano put forth a novel hypothesis in the mid-sixteenth century: frog spawn could be carried up to the clouds by strong gusts of wind and there develop into complete frogs; when fully formed, they rained down in showers. This hypothesis was supported by the British antiquary Robert Plot, who had investigated some reports of frog rains in Staffordshire. In one of these reports, a man walking across a marshland on a foggy morning had his hat covered with small frogs that had fallen on him with the rain. Plot had a spontaneous generation theory of his own: a particular kind of dust could be spread by the wind and fermented by the actions of the sun and the rain, finally developing into frogs. His contemporary Samuel Pepys recorded an interesting conversation with Mr. Elias Ashmole, a founding member of the Royal Society, at a dinner given by the mayor of London. Mr. Ashmole, a firm supporter of spontaneous generation, assured him that "frogs and many other insects" really fell from the sky ready formed. The great Francesco Redi, who had disproved the dogma of spontaneous generation of larger animals, had to find an alternative explanation for the rains of frogs and toads. He proposed, as had Theophrastus of Eresos 2,000 years earlier, that this old superstition was due to a misinterpretation of the fact that, after a heavy rain, a multitude of young frogs left their waterlogged holes and crevices, giving the illusion that they had fallen with the rain. Many eighteenth-century naturalists agreed with him and doubted the phenomenon of amphibian showers. The German herpetologist Dr. Roesel and the British naturalist Gilbert White both denied the occurrence of frogs and toads falling with the rain, as did Carl Linnaeus.

In 1834, a certain Colonel Marmier reported to the French Academy of Sciences that he had, during a ride on a country estate, seen the entire road being covered with small frogs after a heavy rainfall. The old military man did not hesitate to declare that the animals must have fallen with the rain. The leading French herpetologist, Professor André Duméril, criticized this premature conclusion, however, and quoted the observations of Dr. Roesel, who had himself observed the ground being covered with small frogs that emerged from their hiding places among the vegetation after a heavy rainfall. The debate fueled by Colonel Marmier's letter was widely reported in the daily newspapers, and the surprising aftermath was that many Frenchmen sent in reports of their own experiences of showers of frogs or toads. In not

A British shower of small frogs, depicted in Frank Buckland's *Curiosities of Natural History*, with the caption "Suddenly to descend, to the astonishment of rustics." Reproduced by permission of the British Library.

fewer than nine of these, the animals had clearly been seen to fall down with the rain. In 1804, two gentlemen traveling to Toulouse in a charabanc observed a very black cloud, which burst on them. They were showered with an immense number of frogs, which fell on their hats and cloaks and covered the road and surrounding fields. When the charabanc started moving again, the travelers having recovered their senses after this unexpected amphibian deluge, its wheels ran over and killed a large number of these animals.

In another tale, a noblewoman reported that, while hunting, she had been showered with frogs during a violent thunderstorm. A man of science, M. Mauduy, curator of natural history at Portiers, had himself witnessed two falls of frogs during heavy rain, in 1809 and 1822. Probably the most remarkable shower of amphibians in modern times was experienced by the civil servant M. Gayet. In 1794, when he was an army recruit, he had bivouacked near Lalain, along with 150 other soldiers of the grand guard. Nearby, the Austrian army had flooded a large territory with water from the River Scarpe. It was an oppressively hot, sultry summer day. Suddenly, at three o'clock in the afternoon, an enormously heavy rainfall began. The soldiers had to leave the valley in which they had made camp to avoid being submerged by the

Showers of wheat, fish, frogs, and worms; from the *Der Wunder-reiche Überzug Unserer Nider-Welt*, by Erasmus Francisi.

floods. Suddenly, an immense multitude of toads the size of hazelnuts fell with the rain, and began to jump about among the startled guardsmen. M. Gayet spread out his handkerchief and caught a considerable number of toads in it. Some of them had their posterior portion elongated like tails, signifying that they were still in the tadpole state. This heavy shower of toads lasted not less than half an hour, during which time the soldiers could feel the falling animals striking their hats and clothing. After the storm, many of them found toads lodged in the folds of their three-cornered hats; of all the bombardments these soldiers were to suffer during the endless revolutionary and Napoleonic wars, this is likely to have been the one that most impressed them.

In the middle of the nineteenth century, there was a vigorous debate about the reality, and cause, of these bizarre showers of amphibians. Many cases were reported, particularly in France and Britain. In the Yorkshire village of Selby, frogs "about the size of a horse-bean" fell with the rain in 1844. The villagers caught several of them by holding out their hats and noted that the animals seemed remarkably lively after their long wingless flight. This case was published in the *Zoologist,* a reputable scientific magazine, but the editor, Mr. William Newman, added in a postscript that he, personally, was not among the supporters of this "vulgar error" that had gained widespread credence among his countrymen. There were fewer reports of frog or toad showers from the United States. The earliest of them occurred in 1873, in Kansas City, Missouri. A shower of frogs darkened the air and covered the ground for a long distance; this case was published in *Scientific American.*

High-Flying Fish

A medieval shower of fish is said to have occurred in Saxony in A.D. 689, during the reign of King Otto III; it was described by Conradus Lycosthenes, and is the subject of another of his quaint illustrations. According to the *Annals of the Four Masters,* an Irish shower of fish occurred in Tirconnel in 1506. Another early shower of fish was described in the *Philosophical Transactions* of the Royal Society of London. On the Wednesday before Easter 1666, a two-acre field near Wrotham in Kent was found to be covered with small fishes. There were no fish ponds nearby, but there had been a great tempest of thunder and rain some time earlier. The fish, mostly young whitings, were exhibited before the curious in Maidstone, Dartford, and London. Many people had seen the fishes scattered all over the field, and they had no other explanation than that they had rained down from the sky. The early fellows of the Royal Society of London were no strangers to anomalous precipitation: another issue of the *Philosophical Transactions* contained a report of a rain of a substance resembling wheat in Wiltshire, and in yet another volume, the bishop of Cloyne reported some prodigious showers of a greasy, foul-smelling matter in counties Limerick and Tipperary. The country people eagerly collected this substance for use as an ointment for "Scal'd or Sore Heads"; their body odor is unlikely to have benefited from this practice, since the grease was reported by the bishop to have "a strong ill Scent, something like the smell of Church-yards or Graves."

Another early shower of fish occurred in the 1780s, in a town near Paris. After a violent thunderstorm, which had thrown down both trees and buildings, the streets were full of small fish about 4 to 8 inches in length. The common opinion was that the fish must have fallen during the thunderstorm, although several fish ponds nearby were found to be empty of both fish and water. These early observations of numerous fish found outside their natural element did not provide any conclusive proof that they had rained down from the sky, however. This lack of proof made the fish-rains an obvious target for the eighteenth-century rationalists, who altogether denied the existence of rains of live animals. In 1771, little fish had been alleged to fall during a heavy thunderstorm in Cotbus, Germany. A savant named Raphael Eglini described this happening in a scholarly magazine, but when the fish were thoroughly examined, one of them proved to be of a species occurring nowhere near Cotbus. Eglini shamefacedly had to withdraw his former article, and declared the entire incident "incredible."

As for the rains of amphibians, there are also cases in which people had actually observed the high-flying fishes fall down toward Earth. When, in 1809, a large troop of British soldiers was on the march a short distance from Pondicherry in India, they encountered a heavy shower of rain. To the astonishment of all, a great number of small fish fell with the rain, some of them actually landing on the soldiers' hats. These were not flying fish; all were dead, and they descended with no other means than gravity. The commanding officer, General Smith, ordered that some of these fish were to be collected and cooked for his dinner. Another Indian fish rain occurred in 1830 and was described by the naturalist James Prinsep. He himself had not seen the fish fall, but he observed the Indians collecting fish in the fields on the day in question. A fish was actually found in the brass funnel of his own pluviometer. Nine independent witnesses made out affidavits that they had seen fish — some of them rotten, but others quite fresh — fall to the ground with the rain.

What was perhaps the most remarkable shower of fish of all times struck Aberdare in Wales on February 9, 1859. It was raining, with a rather stiff wind from the south-west. While the sawyer John Lewis was busy dragging timber to the saw, he was struck by several small objects falling on his head and back, and even down his neck. He reached under his sweater and shirt, and pulled up several small fish! When looking up, he saw fish falling with the rain all about him. The whole ground was covered with fish, and when Lewis took off his hat, the brim was full of them. The other workmen helped to gather up a whole bucketful of fish, which was emptied into a large pool of rainwater. Both the local and the national newspapers published articles about this "extraordinary shower of fish" striking the heart of Britain. Some debunkers and rationalists objected that a rain of fish was manifestly impossible and contrary to the laws of nature; this kind of phenomenon belonged — along with sea serpent sightings, spontaneous human combustion, and snakes living as parasites in the human stomach — to the "silly season." Fish were as unlikely to rain down from the sky in showers as they were to ride along Regent Street in a coach and four, dressed in tailcoats and top hats. The veracity of poor John Lewis was called into doubt by these skeptics, but the vicar of Aberdare, the Reverend John Griffiths, interceded on his behalf, pointing out that Lewis was a sober, reliable character, and that it was certainly beyond his capability to make up a story such as this. The vicar had recorded Lewis's own testimony and sent it, along with several specimens of the fish, to the famous zoologist Professor Richard Owen. These fish, apparently still living, were exhibited in an aquarium at the Zoological Gardens, Regent's Park. No report on the matter from Professor Owen himself has been kept for posterity. Another savant, Dr. J. E. Gray, of the British Museum, instead entered the debate but hardly did himself proud. He bluntly suggested that the whole thing was a fraud and imposture, and that the other workmen had poured a large basin of fish and water over John Lewis as a coarse practical joke. The Welsh sawyers were probably having a good laugh at the expense of the many gullible journalists and amateur scientists who had been taken in by their prank. This feeble attempt at debunking does not explain the several wheelbarrows full of fish that could be collected in the vicinity of the barnyard, nor that the fall apparently occurred over a considerable area. The editor of the *Zoologist,* Edward Newman, who seems to have been a sworn enemy to all kinds of anomalous precipitation, seconded Dr. Gray's objections; he agreed that a rain of fish was manifestly impossible and that a fraud was the most likely explanation. On the other hand, Dr. Robert Drane, of Cardiff, and the Reverend W. S. Symonds reported to the British Association for the Advancement of Science that they were convinced that a considerable number of fish really had descended with the rain. According to the contemporary newspaper correspondence, both the journalists and the British public took their side against the narrow-minded rationalists. In the 1862 volume of his magazine *All the Year Round,* Charles Dickens commented on the fish and frog rain debate, quoting several of the Indian reports as well as the recent Welsh one, in support of the veracity of John Lewis against the onslaught of the British Museum officials.

Worms from Outer Space

On December 9, 1923, people taking a walk through the cold Scandinavian landscape in Halmstad, Sweden, suddenly observed small dark objects falling down with the snowflakes. They were much frightened to see that these objects were writhing, reddish worms, between 1 and 4 inches in length, falling down like confetti. The police, the fire brigade, and the local newspaper journalists were summoned by the startled observers, but when the constables finally came pounding along, the worms had ceased to fall. The newspaper men wrote lengthy epistles about this unheard-of occurrence; the story later spread to various news agencies, and the worms of Halmstad appeared in the "funny pages" of many foreign newspapers. Several Swedish zoologists were consulted by the journalists, but these worthies seemed to have felt insulted to have been asked such an absurd question. Some believed that the witnesses had hallucinated, others that it was all a tasteless practical joke; but all were unanimous that worms could not fall with the snow. As their dismissive comments were published in various newspapers, new showers of worms occurred on several locations in Sweden, spreading terror and confusion among superstitious people. In early 1924, both Uppsala and Stockholm were bombarded with worms from the sky. A lady walking in a park just outside Stockholm was struck by a violent shower of worms that were crawling about on her hat and fur coat. Now, the zoologists could not deny that something strange was afoot, but they were at a loss to explain this eerie phenomenon. An entomologist suggested that that the worms had crept up onto the newly fallen snow after their lairs had been flooded with water, to give the illusion that they had rained down. This completely disregarded, of course, the many observations of worms falling with the snow. While the men of science were faltering, the Scandinavian crackpots had a field day. Some of them suggested that the worms fell from other planets and demanded that the air force be mobilized to stop this invasion from outer space. Other enthusiasts believed that the worms were no mere space invaders, but the super-intelligent ambassadors of a race of highly developed beings from another galaxy. The worms would not take kindly, they warned, to the brutal Swedes trodding them underfoot instead of inviting them to the royal castle to meet the king and the prime minister.

One of the zoologists pondering these remarkable showers of worms was Dr. Herved Berlin, conservator of the Museum of Zoology at Lund University. After a year of diligent study, he published a booklet that promised to provide the final solution to this mystery. He had found several earlier instances of rains of worms, one occurring in Hungary in 1672, and another in Filipstad, Sweden, in 1749. The latter was described before the Swedish Academy of Sciences, and it is recorded that some specimens of worms were sent to the famous entomologist Baron Charles de Geer. It was not recorded if the academians had had the foresight to mark the envelope "Live animals! Handle with care!," but it nevertheless speaks highly of the eighteenth-century Swedish postal service that Charles de Geer remarked that the worms arrived "alive and in excellent health"! His examination demonstrated that they were black and had six brownish legs, much resembling the specimens described in Hungary 77 years earlier. Herved Berlin procured specimens from several of the Swedish rains of worms, and studied the animals closely. Like those examined by Charles de Geer 175 years earlier, they were compost worms, which normally spent the winter season adhered to leaves in composts and dung heaps. In a series of clever experiments, Berlin demonstrated that when there is much rain, low atmospheric pressure, and generally mild weather, these worms encapsulate quite shallowly in their composts and heaps of leaves. Even if the winter is particularly cold, they remain in this position. If there are some days of mild weather followed by a snowstorm, these worms will sail up into the air using the compost leaves as sails. When they gain sufficient altitude, the freezing cold makes them all release their hold of the leaves, and they fall toward the Earth as a shower. Berlin pointed out that all these three meteorological conditions must be fulfilled if a genuine shower of worms is to occur, which was the case during the extreme Scandinavian winter of 1923 to 1924. This strange epidemic of rains of worms appears to be unique in modern times; nothing even remotely like it has since been observed.

Even Odder Showers

In August 1708, the town of Norrköping, Sweden, experienced a thunderstorm "with a great thundering and continuous lightning, the like of which has scarcely been observed in this country." Just after a house was struck by lightning, with a tremendous thunderclap, a large animal rather resembling a beaver fell from the sky and crash-landed on the high street, just by the pharmacy, with an audible thud. This extraordinary happening was investigated by Bishop Johan Bilberg, who did not doubt that the strange animal really had dropped down from the sky, since several sober and reliable witnesses had actually seen it fall. He was at a loss to explain where it had come from, and how a heavy, wingless beast such as this could have been airborne in the first place. From spiritual reasons, he scornfully rejected the idea of some of his parishioners that it was a flying troll, or even the devil himself, who had inadvertently been struck by lightning while hovering, on mischief bent, over the quiet Scandinavian municipality. The space beaver of Norrköping has lived on, not only in obscure works of Scandinavian mythology but in the vulgar present-day tomes on "the unexplained": the strange animal descending from the sky has been considered a strange and absolutely inexplicable occurrence—and yet another piece of evidence that God, UFOs, or teleportation was at work shifting beavers about in the upper stratosphere.

From a contemporary newspaper report written by a clergyman who was opposed to Bishop Bilberg's theories, it would seem that even this strange and unparalleled happening has a perfectly rational explanation, one that would delight even the most hardened debunker and show that truth is indeed stranger than fiction. It is recorded that, on the fateful evening, some unfastidious individual had kept a capacious tub full of seal's lard and blubber, apparently intended for human consumption, stored on the roof of a tall building. This huge tub also contained the putrid body of a seal cub. When lightning struck the building, the tub of lard was directly hit: it was completely shattered, and its contents were blown sky high. The space beaver was thus instead a flying seal, and the people responsible for its taxonomic

A British shower of wheat, accompanied by a strange apparition in the sky, looking not unlike a UFO.

classification might have been well advised to concentrate on theology, and leave the realm of zoology well alone.

The "Beaver from Hell" is not the only large animal reported to have fallen down from the clouds. The earliest instance was reported by the old Avicenna, who claimed that a calf had once fallen from the sky, possibly from being carried off of a tall mountain by a strong gust of wind. Similar tall tales are not infrequently encountered in the old chronicles of prodigious happenings as well as in nineteenth- and twentieth-century newspaper sources. If various provincial newspapers are to be relied on, showers of lizards, snakes, and salamanders occurred in various parts of the United States during the late nineteenth century. One of these reports was based merely on the finding of some snakes crawling in the main street of Memphis, Tennessee, after a heavy torrent of rain; no animals were actually seen to fall. Some of the other alleged falls of larger animals are even more poorly authenticated, while others, such as a shower of venomous rattlesnakes striking a party of

Irish immigrants in Arizona, seem very much like practical jokes. Even today, reports of large animals falling from the sky can be encountered in the cheap tabloid newspapers, whose priorities are given to reports that Elvis Presley is alive and kicking, that the First Lady was once abducted by aliens, and that UFOs have landed outside the capitol. The revelation that an alligator has fallen from the sky into a suburban garden is treated as a piece of stale old news. At another time, one such paper reported that a dead monkey was found in a garden just outside Washington; it had struck the ground with considerable force, the impetus of the body breaking a solid wooden pole. Some people presumed that the animal had been thrown from an airplane by some sadistic aviators, but airport authorities reported that no identified flying objects had been observed in the area. Instead, there were rumors that a flying saucer had been whizzing past, and that the poor monkey, which had been abducted by evil aliens, had been pushed from their spacecraft without even being given the courtesy of a parachute. A photograph of the dead monkey was reproduced with the caption "Did he fall or was he pushed?"

In September 1945, an old lady telephoned the weather bureau of St. Louis, Missouri, to inquire whether the end of the world had arrived. The reason for her inquiry was that green beans were raining down from the sky just outside her house. The weather expert's reply to this strange query is unknown, but it may well have been to purchase a pair of new spectacles, or even to "take more water with it!" Had he been better read in the annals of odd showers, he could have informed her that, at least according to the newspapers, strange objects from the vegetable kingdom regularly fell from the clouds onto the American countryside. In 1961, green peaches rained down in Louisiana, and the year after, peas fell in Blackstone, Virginia. The British Isles provided two even odder showers: in 1974, eggs fell in Wokingham, and in 1977, nuts rained down near Bristol.

The old German chronicler Conradus Lycosthenes, who did much to preserve records of Europe's strange downfalls to posterity, also described a shower of small crosses that was said to have fallen in Germany in 1503; his illustration of this miracle shows a religious man dashing about to collect the crosses in his tunic. In later years, a strange array of artifacts has been claimed to have fallen from the sky: if Conradus Lycosthenes had known of these occurrences, his outlook on life would have been irrevocably changed. In 1923, fragments of broken porcelain fell during a thunderstorm near Portland, and in 1954, nails fell down in Raritan, New Jersey. In November 1965, a man in Louisville, Kentucky, heard a tremendous explosion coming from his rear garden: when he went out to investigate, the entire patio as well as the surrounding gardens and garage roofs was covered with plastic bags of cakes. It was suspected that they had been dropped from an airplane, but the tabloid press reported that no identified flying objects had been in the area; the police and the meteorologists were completely flummoxed. Four years later, golf balls fell from the sky during a thunderstorm in Punta Gorda, Florida, and hundreds of them rolled about in the street. After a week of investigation, the local police department declared that unnatural freak accidents such as this were outside their field of expertise: the identity of the heavenly golfer remained unknown.

Several old chronicles mention showers of silver or money, and it is curious to note that similar occurrences have been reported in the twentieth century. In 1940, silver coins worth several hundred kopeks fell during a storm in the Gorkij area of the former Soviet Union. In January 1976, 10-mark notes descended from a clear sky over Limburg, Germany, and two observant clergymen managed to gather up more than 2,000 marks worth of this money from heaven. Another shower of money, although coins of lower denominations only, puzzled the parishioners of St. Elizabeth's Church, Reddish (between Manchester and Stockport) in May 1981. Pennies were heard to tingle on the gravel of the churchyard, and a girl claimed to have seen a 50-pence coin fall from nowhere in front of her. The children found many pounds worth of copper and silver coins, which led to a roaring trade in the local sweet shops. The rector heard news of this and suspected that the children had raided the church poor box, but he was convinced by their tall tale of pennies from heaven when he found over 2 pound's worth of coins in his own churchyard. The investigation could provide no explanation whatsoever, although the clergymen seem to have spent much time searching for magpies' nests rather than considering the possibility of a cunning practical joker.

Manna and Letters of Condolence from Above

In the late nineteenth century, progress in natural history began to catch up with the odd showers of olden times, and several phenomena that had seemed strange and inexplicable 200 years earlier now received a rational explanation. For example, the Biblical rains of "manna" may well be based on actual observation. In several instances, travelers in the desert parts of Egypt, Persia and Turkey had found large amounts of light, whitish, round objects about the size of a hazelnut. They are quite edible and can be used in baking bread. Well into the nineteenth century, the local inhabitants of these parts commonly believed that this manna had fallen from heaven, but modern botanists have provided a rational explanation of this phenomenon. The substance known as manna consists of lichens of the genus *Lecanora,* which sometimes occur in great numbers. They may easily be blown away over the surface of the ground. They tend to accumulate in depressions and are easily drifted into masses during the runoff of rainwater. The hungry desert traveler finding them is likely to fall on his knees and praise the Lord, just like the Children of Israel.

Modern scientists have also been able to explain the rains of blood, which were considered as dreadful portents throughout antiquity and the middle ages. A shower of blood fell at the death of Julius Caesar, and blood rains foreboded the war against Carthago, the death of Hadrian II, and the marauding of Attila the Hun. Gregory of Tours told of a rain of blood that fell over the city of Paris in 582, during which many people had their clothes stained with blood and cast them off in horror. In 1648, a heavy shower of blood fell in Mecklenburg, and a voice from above called out "Wehe! Wehe!" In 1653, according to the nonconformist minister Samuel Clarke's strange *Mirror or Looking Glass Both for Saints and for Sinners,* a black cloud was seen over the town of Poole in 1653; it was dissolved into a shower of blood that fell warm on people's hands. Some green leaves spotted with drops of blood were sent to learned men in London. These red rains are due to the air being charged with scirocco dust following sandstorms in the Sahara desert; this dust may, during certain meteorological conditions, color the rain a bright red. Another eerie phenomenon of medieval times was the yellow "rains of sulphur and brimstone"; similar anomalous precipitation has occasionally been observed during the nineteenth and twentieth centuries; the explanation is that the rain is charged with huge amounts of pollen from nearby pine woods.

One of the most remarkable downfalls in all times occurred on January 31, 1687, near the village of Dauden in Courland. Large amounts of a blackened, paper-like material fell during a storm. This Trauer-papier, or mourning paper, was by many considered a dire portent that the end of the world was near. Some people imagined that there was writing on the paper, and racked their brains to find out what was written on these letters of condolence from the sky. After several years, without any cataclysm occurring, they believed themselves secure, and the black-edged paper was deposited in a museum. When, 150 years later, this mourning paper was examined by the German scholar C. G. Ehrenberg, it was found to consist of dried algae, which had assumed a paper-like texture. This substance was light enough to be carried a considerable distance by the wind.

Do Little Fishes Fall from the Sky with Rain?

At the advent of the twentieth century, the majority of zoologists and weather experts were still unwilling to accept the possibility of rains of fish, amphibians, and other animals. The early cases were dismissed as the result of superstition, and the later ones either as hoaxes or as misinterpretations of floodings or mass migrations of amphibians by imaginative laymen. Dr. E. W. Gudger, associate and bibliographer in ichthyology at the American Museum of Natural History, worked for many years to produce a *Bibliography of Fishes,* covering all aspects of the natural history of these animals. He was puzzled to encounter many reports of rains of fish in the older literature and decided to publish a review on this subject, detailing not fewer than 48 cases, from A.D. 300 to 1901. It began with the words: "Do fishes fall in rains from the sky? To this question both the layman and the scientist are well-nigh unanimous in giving a negative answer." Dr. Gudger was somewhat wary of the reaction of his professional brethren to his new bizarre undertaking, since at least one of them had expressed himself in strong language on being asked such a bizarre question. Dr. Gudger was probably relieved to note that his article, published in *Natural History* in 1921, was favorably received; indeed, many people wrote to him to add further cases, or even to tell him about their own experience of high-flying fish. In two supplements to his original article, Dr. Gudger was able to describe 23 additional fish rains. It is rather strange that he never discussed rains of frogs and toads, or, indeed, any kind of anomalous precipitation other than showers of fishes; perhaps he had made it a rule, as an ichthyologist, never to venture outside his own field of expertise.

In 1946, Mr. Bergen Evans, Professor of English at Northwestern University, ridiculed the old belief in rains of blood, milk, frogs, and little fishes in a popular book entitled the *Natural History of Nonsense,* and also in several magazine articles. His book is an amusing overview of various modern myths

A shower of cats, dogs, and pitchforks: an amusing caricature drawing by George Cruikshank. From the author's collection.

and "vulgar errors," but its author's lack of knowledge in medicine and zoology did not impede him from expressing decided opinions on matters within these fields — and his passion for debunking sometimes seems to have got the better of him. Bergen Evans read only the first of E. W. Gudger's articles but boldly declared himself to be unimpressed by its reasoning, and implied that this obscure boffin's imagination must have had run away with him. Dr. Gudger answered him in *Science* magazine, pointing out that he had collected 78 cases of rains of fish alone, some of them described by reliable naturalists. Others cases had been observed by several independent witnesses, often simple, uneducated people who knew nothing about the fish rain debate. Bergen Evans was adamant, however. He ridiculed Dr. Gudger's claim that some cases were "scientifically attested," declaring that the whole thing was manifestly impossible, and that rains of little fishes was nothing more than a silly old wives' tale. It would take an expert zoologist's personally witnessing that fishes had fallen from a sky, free from birds or other objects, and local air traffic controllers' signing an affidavit that the air space was free of planes on the time in question, before he was prepared even to consider the question. None of Dr. Gudger's cases passed these rigorous criteria, but in

1949, the scientist Dr. A. D. Bajkov reported, in a letter to *Science,* a rain of fish which he had personally witnessed, in Marksville, Louisiana. The people in town were very much excited by this strange phenomenon. The bank cashier and two merchants were struck by falling fish, and many other people saw them fall with the rain, along a corridor 1,000 feet long and 75 or 80 feet wide. Dr. Bajkov noted that the day before, numerous small tornadoes or "devil dusters" had been observed in the area.

A thorough review of the annals of anomalous precipitation would support Dr. Gudger's viewpoint: between the years 687 and 1985, not fewer than 117 probable or verified showers of fish occurred, and in 36 of them, the fish were clearly seen to fall through the air. During the same period, there are also records of 58 showers of frogs or toads; in 39 of these, the animals were observed to descend from the clouds. These showers of fish and amphibians have been observed all over the world. Quite a few have occurred in the United States, India, Ceylon, and Malaysia. Britain and France were also among the preferred targets, while there are fewer instances reported from central Europe. Contrary to the predictions of Olaus Magnus, there is also a marked scarcity of Scandinavian odd showers, the strange epidemic of worm showers of 1923 to 1924 excepted.

The intensity of these anomalous showers of fish or amphibians has varied greatly: sometimes, only one or two animals have fallen from the clouds, while on other occasions, they have resembled the plague of raining frogs striking the cities of Peonia and Dardania, or the tremendous shower of toads witnessed by M. Gayet in 1794. On July 2, 1901, frogs and toads fell over Minneapolis during a violent rainstorm: according to the newspaper reports, certain streets and sidewalks were completely filled with "quackers," and neither people nor cart horses could wade through the mass of squirming bodies. After a rain of eels outside Coalburg, Alabama, many farmers brought carts to take away large loads of dead eels, which they used as a fertilizer in the fields. In 1969, a lady living in Buckinghamshire was just leaving home to go to a party when it suddenly started to rain violently. The doors and windows of her house had been left open, and hundreds — even thousands — of frogs poured in with the rain. Only with great difficulty could she clear the house of the invading amphibians. When she reached the party, many people doubted her tale of the frog shower, but she silenced them by picking up a couple of frogs that had got stuck in her baggy trousers.

The eccentric American journalist Charles Fort spent 27 years at the New York Public Library and at the British Library's round reading room, scanning the files of many periodicals, newspapers, and popular science magazines for articles and reports of various strange and macabre phenomena. He collected early UFO observations, reports of ball lightning and spontaneous human combustion, and appearances of ghosts, ghouls, and poltergeists. Quite aptly, he called these phenomena "damned," since they were sneered at or ignored by the scientists of his time. Instead of classifying his observations, Charles Fort presented them in veritable disarray. Being a "phenomenalist," he argued that anything might be possible in nature, and that all attempts at interpretation were futile. An amazing number and variety of tales about anomalous precipitation are scattered throughout his books: rains of frogs, fish, fungi, stones, hatchets, masks, and the "ceremonial regalia" of savages. His own explanation of this occurrences was that what he called a "Super-Sargasso Sea" hovered over planet Earth, and that live animals and other substances regularly rained down from it.

For modern devotees of paranormal phenomena, Charles Fort has become a cult figure: both in Britain and in the United States, there are societies devoted to the continuation of his life's works. His theory about the Super-Sargasso sea has proved a trifle risqué for even the most audacious modern "crypto-biologists": no aviators or astronauts have reported that their planes or rockets have encountered such an anomalous ocean. The modern Forteans instead suggest that mysterious forces sometimes teleport animals from the face of the earth up among the clouds, and that they later fall down with the rain on some other location. Using the most absurd reasoning, several "occult philosophers" have added their own embellishments to this teleportation theory; it would have dismayed Carl Linnaeus that even the tall tale about the falling lemmings of Lapland has been uncritically repeated, even in books published in the 1980s! The Forteans have not provided any explanation, however, why only small animals, such as frogs and fish, are affected by this strange uplifting power. There are no reports of flying pigs and cows buzzing about, nor has any rhinoceros fallen through the roof of a suburban sitting room!

In view of the compelling evidence from both historical and modern cases, it cannot be denied that little fishes, frogs, and toads sometimes fall down in showers. Although some cases can be explained as hoaxes, ill-researched or fraudulent newspaper stories, or misinterpretation of mass migrations of amphibians, the vast majority are beyond reproach. There is also a quite natural explanation of these odd showers, which, unlike the musings of the Forteans, not only fits all the facts but also keeps within the established laws of nature. A waterspout is a pillar of water drawn up by a powerful whirlwind; it can suck up vast amounts of water, mud, and earth from the ground. Some waterspouts resemble a giant hourglass, with one end traversing the earth and the other being in contact with a large cumulus cloud. It has several times been observed that waterspouts can transfer large quantities of sea or lake water up toward the cloud; the amphibians or little fish in this water are also taken for a ride in the huge merry-go-round, whirling round in the waterspout with tremendous speed. In 1836, a report of whirlwinds accompanying a storm over Florence mentioned that among various larger objects spinning in the air, several large fish, apparently drawn out from the river Arno, could be discerned. Sometimes, the fish and amphibians travel ten, or even twenty miles from their natural habitat before being precipitated with the rain; the hardy animals are often still alive when they hit the ground. On some occasions, the low temperature froze the fish stiff, however, and they were shattered into fragments when they hit hard objects on the ground—a bizarre sight indeed! It has even been observed that an enormous hailstone has contained a frog or a little fish. In 1896, a large hailstone that fell in Essen, Germany, contained a crucian carp, 1½ inches in length; two years earlier, a gopher turtle 6 by 8 inches in size was stated to have fallen with the hail in Vicksburg, Mississippi, entirely encased in ice.

The famous French naturalist Count de Castelnau, who examined a fish rain occurring near Singapore in 1861, suggested to his incredulous colleagues that a whirlwind or tornado might have caused this phenomenon. The greatest modern authority on fish showers, the American naturalist Dr. E. W. Gudger, also advocated the whirlwind theory. During the 1980s and 1990s, British zoologists and meteorologists belonging to the Tornado and Storm Research Organization have put forth strong evidence that the showers of fish and amphibians may well be explained without resorting to supernatural phenomena. The meteorologists have even at times been able to reconstruct the path of the whirlwinds and the site where the fishes were drawn up into the cloud. What is strange, and hitherto unexplained, is how such an enormous amount of toads or fish could fall from the sky; in some instances, the whirlwind must have emptied an entire lake or bog of its fish or amphibian population. Nor has it been explained why the falling animals are seldom accompanied by mud, algae, and other debris; furthermore, it has many times been remarked that the falling frogs are of the same approximate size. It is possible that the animals are separated by some obscure kind of "air chromatography," making big toads fall before smaller specimens.

Clouds of Frogs over London

In the old chronicles, the British Isles were stated to have been frequently bombarded by strange objects from above, and this has been continued well into our times. London itself has several times been targeted by these anomalous downfalls. The *Mirror* of August 4, 1838, contains an article with the promising heading "A Shower of Frogs in London!" The facts themselves are less impressive: a man strolling down Tower Street saw some dozens of young frogs hopping about on the foot and carriage pavements, and naively presumed that they must have been precipitated to the earth in a heavy shower of rain that had fallen about a hour earlier. Another alleged shower of frogs struck the garden of the zoologist Edward Jesse. One morning in the early 1850s, this gentleman went for a walk in his walled Fulham garden. He was much surprised to find his garden full of little frogs, jumping merrily about. The day before, it had been wholly devoid of amphibians, and the soil was dry gravel, with no moist spot in which spawn could have been deposited. The garden was walled, like many others in this area, and he could see no apparent way for the animals to enter it except through the house—and from the sky. This curious experience converted Edward Jesse from being a skeptic to keeping an open mind with regard to odd showers. I myself have many times walked in my walled Fulham garden after rain but have never found anything more interesting than garden slugs! London seems to have been spared from further anomalous downpours until 1984. While watching television, Mr. Ronald Langton, of East Ham, heard some heavy slapping sounds from the roof. The morning after, several flounders and smelts were found in the garden and on the roof. In nearby Canning Town, fish were found in at least two gardens. Although there had been a heavy thunderstorm in the night, no waterspouts had been observed. The fish were taken to the Natural History Museum, where the ichthyologists declared them to be just what one might find in the Thames below Newham. A skeptic pointed out that there was a large heronry at Barking, and that a flight of herons returning from the Thames might have been attacked by crows, thereby being forced to disgorge their catch over East Ham. This may well have been the proper explanation.

It is difficult to gauge the impression made by a shower of anomalous precipitation striking the heart of London. A shower of "quackers" pattering against the umbrellas of a crowd of travelers at Sloane Square or Fulham Broadway would naturally stir up some curiosity, although it is unlikely that mass hysteria would break out; the Londoners would probably chase the toads along the winding streets, in apparent danger of being run over by motorcars, scooters, and taxicabs, catching the amphibians in bags and bowlerhats to bring them to the Natural History Museum. Things were very much different when a small American town, situated in the thick of the notorious Bible Belt, was struck by a brisk shower of fish in the mid-1920s. According to the (probably exaggerated) newspaper reports, people ran wildly along the streets, calling out "It's raining fish!" The roof of the Baptist church was covered with fish: inside, the congregation was kneeling, confessing their sins before the Last Judgment. Some religious men ran into the haberdashery to buy expensive suits and tailcoats, to be well-dressed before Resurrection; others instead bought expensive coffins, lay down in them, and slammed the lid shut. In midst of all this confusion, some individuals, probably hard-necked rationalists, could be seen gathering the fish from the sky into their corfs and frying pans, no doubt planning to cook them for their dinner.

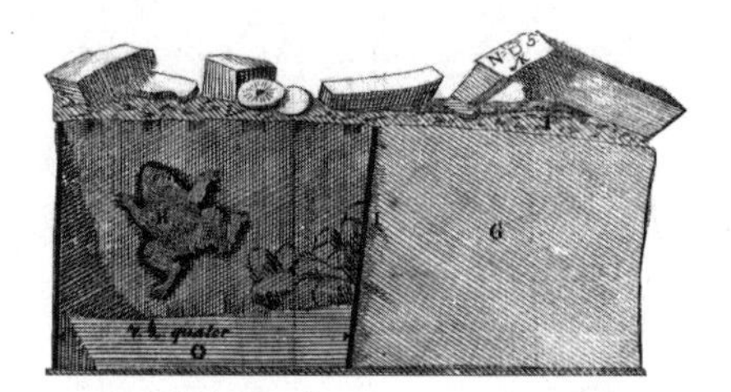

Toad in the Hole

❧ *Let dotards with tenacious force*
Cling to this waning planet—
I'd rather soar to death's abode
On eagle's wings, than live like a toad
Pent in a block of granite.

James Smith, *Chigwell Re-visited*

ON MAY 8, 1733, THE MASTER BUILDER MR. JOHAN Gråberg went to the quarry of Nybro, situated near the village Wamlingebo on the Swedish island of Gotland. In this quarry, he was supervising the cutting of a shipload of large stone boulders that were to be used for the building of the new Royal Castle in Stockholm. The work proceeded in good order, and the master builder was thinking of his luncheon, when two of the quarrymen, Anders Halfwarder and Olof Sigräfwer, suddenly came running up from the excavation, both in a most excited state of mind. While cutting large blocks of sandstone more than 10 feet below the level of the earth, Anders Halfwarder had been frightened out of his mind when he saw *a large frog* sitting in the middle of a large boulder that he had just cut in two with his hammer and wedge. Reluctantly, Mr. Gråberg followed the superstitious workmen down into the quarry, where he, too, was greatly startled by the sight of the frog sitting inside the boulder. Unfortunately, the part of the stone nearest to the frog was so porous that the violence of the blow had fragmented it, and the impression of the animal's body was destroyed. The frog was in a lethargic state, and Mr. Gråberg could not provoke it to move, even when he lifted it out on a spade. Its color was a grayish black, with some speckles on the back; under the belly, its color was paler. When Mr. Gråberg touched its head with a stick, it closed its eyes. Its mouth was closed and covered with a yellow membrane. The master builder's examination of the mysterious stone frog, which was regarded with superstitious awe by the workmen, was cut short by his impatience: for some reason, the brutal Swede wantonly beat it to death with his heavy shovel. The quarrymen put the flattened body of the supernatural animal on a polished slab of stone, and laid it in state in their cabin.

The Stone Frog Before the Academy of Sciences

Later that afternoon, when Mr. Gråberg was leaving the Burgsvik quarry, he was struck by belated qualms of conscience "for being the Slayer of that extraordinary Animal, that might have lived for many hundreds of years within its stony Prison." He returned to the quarry and recovered the frog's corpse, which he took with him to Stockholm. Six quarrymen and farmers, who had been present when the supernatural animal was freed from its prison within the stone, all signed a certificate of the truth of this extraordinary happening, and the Parson of Wamlingebo assured that all six were "sturdy and reliable men." Some days after the stone frog had been discovered, the quarryman Anders Halfwarder fell badly ill, and Mr. Gråberg wrote that all the workmen believed that the frog, or rather the malignant mountain spirit inhabiting it, had poisoned him.

When Mr. Gråberg came to Stockholm, he told several scholars about his remarkable experience. One of them, a provincial medical practitioner named Dr. Johan Pihl, imagined himself to be an authority on natural history. After analyzing the case, he stated his belief that "frog-spawn had entered the stone in some way" and gradually developed into a fully grown frog during a period of many years. Dr. Pihl's great ambition was to become a fellow of the Swedish Academy of Sciences, and he wrote a treatise about the strange stone frog, consisting of thirty-nine closely written pages, which was submitted for the academy's scrutiny. Unfortunately for Dr. Pihl, the referees were wholly unimpressed by his lengthy theorizing about these matters. At the academy's meeting of November 14, 1741, it was decided that the doc-

A drawing of Mr. Gråberg's frog and a cross-section of the quarry in which it was found. From the *Kungliga Vetenskaps-Akademins Handlingar* of 1733.

tor's paper was to be put in the archives, where it is still kept today. Only the brief section containing Mr. Gråberg's account of the discovery of the frog was published in the academy's *Transactions.* It was illustrated by an excellent engraving showing the upper and lower aspect of the frog's cadaver, and a transverse section of the quarry where it was found. It does not seem as if the frog's cadaver was dissected, or even examined, by any naturalist of repute; Gråberg's naive statement that "within its Skin, there did not seem to be any Innards, although the Frog, while she lived, had seemed reasonably thick" does not inspire confidence in his knowledge in amphibian anatomy. Indeed, from the existing account and illustration, the animal might just as well have been a toad.

Dr. Pihl felt deeply hurt that his learned treatise was refused in favor of the simple-minded builder's notes on the subject, and he made no further attempts to become a fellow of this ungrateful academy. He had likely hoped that Carl Linnaeus would have been a referee of his paper, since this great biologist shared his belief in the these tenacious underground amphibians. In one of his manuscripts in Swedish, *Beskrifning Öfwer Stenriket,* written from 1747 to 1748, Linnaeus stated his belief that frogs and toads could live for centuries enclosed in sandstone. This variety of stone was remarkably porous and was filled with water, from which the silent, immobilized stone frog obtained its frugal nourishment. Carl Linnaeus also claimed that frogs had an almost supernatural capacity to survive without contact with fresh air. There was a rumor in Stockholm at the time that Linnaeus himself had examined Mr. Gråberg's remarkable stone frog, but this cannot be substantiated from Linnaeus's published writings. In the 1760s, after the death of Mr. Gråberg, his frog is recorded to have been kept in the natural history collection of Count Carl Gustaf Tessin at Åkerö Castle; it was described, in an extant manuscript, by the lecturer in chemistry of the University of Uppsala, Mr. Philip Tidström. No trace survives about the stone frog's whereabouts today.

Toads and Frogs in Stone

Mr. Gråberg's paper on the stone frog was originally written in Swedish but, soon after its publication, was translated into German, Dutch, French, and Latin. It arose much interest in this obscure phenomenon among European scholars and natural historians; many of them reviewed other, older cases of entombed frogs and toads from the literature, or made experiments to determine the longevity of toads enclosed in narrow stone containers. Using a modern expression, the toad in the hole became a priority field of research during the second part of the eighteenth century. This did not happen through the infusion of millions of government grants into laboratories and research institutes, but because the scholars of the Age of Improvement, who believed themselves capable of explaining all obscure phenomena in natural history, were unwilling to accept the presumed immortality of these elusive frogs and toads.

In the old chronicles of monsters and marvels, the eighteenth-century investigators found several ancient tales resembling that of the Swedish master builder. In William of Newburgh's *Historia Anglia,* it was told that, in 1186, a large stone was found. Since it seemed to comprise two stones joined by some adhesive matter, a bishop ordered it to be split; when it was cloven in two, a living toad with a gold chain around its neck was sitting in its center. Everyone understood that this could be nothing but sorcery and witchcraft, and the stone was buried intact without anyone being intrepid enough to divest the toad of its ornament. A much more reliable observation was reported by the old chronicler Fulgosius: in his treatise *de Mirabilius,* published in 1509, he briefly reported that once, at Autun, several people had observed a fat toad being found inside a stone.

The famous surgeon Ambroise Paré supplied details of another French toad in the hole. In 1575, when he was in his vineyard near Meudon, he had ordered some workmen to break up a couple of large stones. In the middle of one was found a large living toad. Paré was greatly astounded, since the stone had no visible opening to the outside. He wondered how the animal had been born and how it had been able to grow and stay alive within the block of stone, but the quarryman, who was apparently a greater authority on toads in the hole than the Founder of French surgery, said that this was not the first time he had encountered toads and other animals in the center of stones. Ambroise Paré concluded that the entombed toads must have been formed through spontaneous generation: some humid matter within the stone had putrefied to produce these animals.

Some other sixteenth-century writers believed that the entombed toads were of demonic origin; they were cousins of the basilisk and almost as poisonous as this monstrous being. A seventeenth-century German mining manual advised the workers to leave the mine posthaste if a subterranean toad was discovered: these supernatural animals were extremely poisonous, and their breath was enough to kill a man. The mineralogist Georg Agricola wrote that the German millers used to scrutinize the blocks of stone that were to be made into millstones, to avoid those that contained a toad in the hole. When the millstones ground, the animal would become hot and spew forth its poison into the grain, thus becoming a mass murderer, to the detriment of the miller's future reputation. A most remarkable occurrence was reported by another German scholar, Dr. P. J. Sachs, in his *Gammarologia Curiosa*. In 1664, a friend of his, Count Hermann of Gleischen and Hatzfeld, had visited the castle of Count Fürstenberg near Cologne. This nobleman had in his possession a round stone, which contained a living frog. When the stone was lifted up, the frog croaked loudly; this strange, hollow sound was likened to the chorus of frogs in the play of Aristophanes: *Koax brekekex!* When finally, the stone was broken, the frog jumped out alive.

Professor Robert Plot, F.R.S., the first keeper of the Ashmolean Museum at Oxford, was greatly interested in the subject of entombed toads. In his *Natural History of Staffordshire,* published in 1687, he described not fewer than three cases of toad in the hole. One of them had been described in a letter from Dr. Pierce, a Bath physician, to Mr. William Musgrave, secretary of the Philosophical Society of Oxford. A large limestone had been put as a stepping stone for passengers in the middle of a cartway over two rills. People had been puzzled by a loud croaking sound coming from this stone. At length, it was resolved that the stone was to be broken open: "in a cavity near the middle a large Toad was found as Bigg as a man's fist, which hop't about as briskly, as if it had been bread in a larger room." An even more astounding case came from Statfold, where the top stone of the spire of the churchtower had fallen down and broken: "There appear'd a Living Toad in the Center of it, which (as most of the rest are said to doe) dyed quickly after it was exposed to the Air."

The records of Chillingham Castle give details about another famous seventeenth-century English toad in the hole. Legend has it that when a large block of stone was sawed into a huge chimneypiece, which was to be put in the spacious entrance hall of this castle, a cavity containing a living toad was found. To celebrate this occurrence, a large heraldic device was painted, displaying a large toad on a shield, held up by two frogs, with the motto "Est et a Jove Bufo." A Latin inscription was made, allegedly by the learned bishop of Durham, which would explain a reference to William Harvey's novel work on generation in second part of the poem, and place the discovery of the toad in the 1660s or early 1670s. The poem contains a challenge to Aristotle (the Stagyrite), since the toad in the hole was a greater marvel than the tides of the strait of Euripus, which had puzzled Aristotle. It was translated into English by Lord Ravensworth:

Ho, Stagyrite!
If you wish something more wonderful than your Euripus,
Come hither!
Let the tides flow and ebb, and be he lunatic
Who robs Trivia of her due honour.
Lo, for you something novel, which Africa bears not,
Nor Nile on his sandy shores.
To wit, fire and pure flame,
Yet without vital air.
Out of the dark recess of the split rock,
As much as you see, the hands
Of the midwife stonecutter gave light
To a living toad.

In 1858, a writer in the *Archeaologia Aeliana* could well remember having seen this chimneypiece with the toad's cavity, and another antiquary, writing in 1884, was able to inspect chimneypiece, painting, and inscription. According to the present-day custodians of Chillingham Castle, they are still there.

One of the most influential mid-eighteenth-century writers on the prob-

The toad in the hole is discovered: a rather fanciful illustration to Philip Henry Gosse's *Romance of Natural History*.

lem of the toad in the hole was the Frenchman Claude Nicolas Le Cat, whose thesis on this subject was laid before the Academy of Sciences at Rouen in 1762. A small toad had been found in a rock near Écretteville outside Rouen, and some local dignitaries instructed M. Le Cat, who was a noted surgeon and naturalist, to inquire into these matters. Like many contemporary scholars, he disregarded the spontaneous generation hypothesis. Nor did he believe in another notion, coming from some conservative French academics, that toads created by God at the beginning of time had accidentally been encased in stone. M. Le Cat believed it to be a general law of nature that a sober, frugal, and inactive manner of living could prolong the individual's life to a remarkable degree: a man could even live to be several hundred years old if he was imprisoned in a narrow cell and denied access to alcoholic beverages, tobacco, and unhealthy, rich food. He estimated that the toad in the hole was about 3,000 years old. He presumed that it had entered the stone after its spawn had trickled down a crevice in the rock, but he did not theorize further as to how this could have been accomplished.

The majority of seventeenth- and eighteenth-century reports of a toad in the hole were of English origin. Other reports hailed from the European continent, mainly France and Germany, and a few were from more exotic locations; one of Le Cat's cases was discovered at Guadaloupe. An American toad in the hole was described by the plantation owner Mr. Samuel Peters of Hebron, in the Colony of Connecticut. In 1770, this gentleman had become increasingly annoyed with a large stone boulder that had been situated in the high road opposite his house for more than 150 years; by the wear and tear of the road, it had gradually grown quite high and had become a nuisance to carriages. Mr. Peters decided to do something about this. He ordered his slaves to dig a wide ditch around the stone, and a miner perforated it with an auger. The hole was charged with gunpowder, which was fired off after the spectators had retired to a safe distance. After the tremendous explosion, chips of stone flew about, but two large fragments of the rock still lay inside the ditch. One of them had a small orifice, a circumstance that puzzled Mr. Peters. When the men widened this opening, they were astounded to see a cavity as large as a goose egg, in which lay a frog that almost completely filled it. The animal was bleeding from one leg but otherwise seemed unaffected by being so rudely awakened from its stupor. Mr. Peters exhibited the strange stone frog before the curious for many weeks.

A certain Mr. G. Dodsley, writing a book of fables in imitation of Aesop in 1761, included a story titled "The Toad and the Ephemeron." A toad in the hole, freed by some workmen, sat swelling and bloating, and bragged that it was "a specimen of the antediluvian race of animals. I was begotten before the flood. And who is there among the present upstart race of mortals that shall dare to contend with me in nobility of birth, or dignity of character?" An ephemeron that was flying by replied that the vain boaster was "as insensible as the block in which thou wast bred." Itself, it had enjoyed the warmth of the sun and the light of the day: "My whole life, 'tis true, is but of twelve hours; but even one hour of it is to be preferred to a thousand years of mere existence, which have been spent, like thine, in sloth, ignorance and stupidity." John Wesley was inspired by the toad in the hole to make a similar analogy. In a sermon preached in Rotherham in July 1790, he compared the empty, valueless lives of men living without God with the trance-like state of the entombed toad. Edward Bulwer Lytton, a celebrated writer of sensation novels in his day, made quite another use of the entombed toad metaphor in one of his short stories, "A Manuscript found in a Mad-house," which was published in 1847. In this highly strung Gothic tale, a hideously deformed boy laments his lonely, wretched childhood: "I was like the reptile whose prison is the stone's heart—immured in the eternal penthouse of a solitude to which breath of friendship never came—girded with a wall of barrenness and flint, and doomed to vegetate and fatten on my own suffocating and poisoned meditations."

Bizarre Experiments

In September 1770, a live toad was found inside a stone wall at the castle of Le Raincy in France. This once more stirred the interest in the subterranean amphibians, and there was much speculation about this subject. M. Jean Guéttard, a fellow of the French Academy of Sciences, made a long speech before this academy, describing the toad in the hole as one of the most puzzling enigmas in natural history: he urged his fellow academians to spare no labor to solve this mystery, which had baffled the naturalists for more than 200 years. His colleague M. Louis Hérissant was inspired to perform an ambitious series of experiments to test the capacity of toads to withstand hunger, thirst, and suffocation. Three fully grown toads were put in boxes, which were sealed with plaster in the presence of several academians. Eighteen months later, they were opened, and two of the toads were seen to be alive. The boxes were sealed and put away a second time, but M. Hérissant was not destined to see the result of his bizarre experiment: he died in October 1773 and was survived by the toads. In his eulogy on M. Hérissant, read before the academy in 1777, Jean Guéttard mentioned that this gentleman, apparently a scientist to the last, had left him the boxes of toads in his will, with instructions to open them some time after his death. This was done, just as prescribed, but the toads were all dead and desiccated.

From the mid-1700s onward, many English clergymen and amateur natural historians were eager to put the viability of toads to the test. Usually, they put a toad in a flowerpot, sealed it with plaster or mortar, and buried it in their gardens. After waiting for some time, the pot was unearthed and the toad freed. Usually, in these experiments, the animal turned out to be alive and in good health. In one of his unpublished manuscripts, the naturalist Gilbert White described an experiment made by an Oxford student who buried a toad in a garden pot for 13 months; when dug up, the animal was alive and well, and "considerably grown." Even the zoologist Edward Jesse's toad, which had been closed up in a flowerpot for 20 years, jumped out with vigor when the pot was opened. These experiments were often widely publicized in the local newspapers, and it was concluded that toads could live forever if left alone in a small cavity without anything to eat or drink.

Dr. William Buckland, the professor of geology at Oxford, and later dean of Westminster, became interested in the entombed toad phenomenon when his friend Luke Howard suggested that the existence of the toad in the hole might provide evidence for or against the gradualist and catastrophic theories of the origin of the Earth. Buckland asked several scientists of his acquaintance how long they believed a toad could survive imprisoned inside a rock, but since their estimates ranged from a few days to 4,000 years, he rightly decided that they knew as little of this subject as he did himself. Buckland did not rely on the primitive experiments with toads in flowerpots, which were often carelessly done. Sometimes, the pots were imperfectly sealed, and in other instances they had corroded, enabling moisture and insects to enter the toad's enclosure. In November 1825, Buckland himself planned a rigorous set of experiments to determine the reality of the toad-in-the-hole phenomenon once and for all.

In a block of coarse oolithic limestone, twelve circular cells were prepared, each about 1 foot deep and 5 inches in diameter. Six large toads and six smaller ones were put in these holes. The cells all were sealed with circular plates of glass, which were fitted into a groove in the stone and sealed by a luting of soft clay. Twelve smaller holes were prepared in a block of siliceous compact sandstone, manned with toads and sealed with glass plates in the same manner. A double cover of glass and clay was placed over each block of stone, and the blocks were buried in Professor Buckland's garden under 3 feet of earth. In December 1826—1 year later—the blocks of stone were dug up and examined. All the toads in the sandstone block were dead and decayed. All the small toads in the limestone block were also dead. To his amazement, Buckland found that the larger toads in the porous limestone block were still alive, and two of them had even gained weight. The block of stone was resealed, and the mute prisoners buried a second time. Buckland examined them several times during the second year, to see if they hibernated, but this was not the case; they all were awake, sitting in their cells, but their emaciation increased each time the stone was unearthed, until finally, they all were dead.

The results of William Buckland's experiment made him doubt the reality of the toad-in-the-hole phenomenon. It was now proved that these animals could not survive in the compact sandstone, which did not admit air to their cells. Even in the porous limestone, even the best-nourished toads starved to death within 2 years. The majority of Buckland's contemporaries agreed with his conclusions, particularly as the French zoologists Duméril and Edwards had performed a series of similar experiments, with results much resembling those of Buckland. The major part of the zoological establishment had thus permanently lost faith in the toad-in-the-hole phenomenon; even today, zoologists quote William Buckland's experiments as the strongest argument against the existence of these subterranean amphibians. It is interesting to note that the stones used in Buckland's experiments were made by masons employed to work on the old residence attached to the canonry. The house that he occupied at the time has since been assigned to the Archdeacon of Oxford, and according to the *Life and Correspondence of William Buckland DD FRS,* by Elizabeth Gordon, the Archdeacon Palmer, who resided there in 1894, placed the stones in his garden in memory of the toad experiments.

The immediate reflection, when surveying these bizarre experiments, is that it is a disgusting, not to say barbaric, practice to immure living animals in this way. Such qualms did not occur to the early–nineteenth century vivisectors, who freely used both toads and higher animals as "cannon-fodder" in their cruel experiments, but ethical concerns of course prevent these investigations from being repeated today. The most competently carried out, and clearly described, of these experimental immurements of toads were those performed by William Buckland. In spite of his careful preparations, even these experiments have certain important flaws, and there is good reason to doubt whether they really disprove the toad-in-the-hole legend. William Buckland had clearly been collecting toads for some time, and at least some of his specimens may just have come out of hibernation, and thus not have been strong. He also disregarded the fact that the toad's metabolism and disposition to hibernate is temperature dependent; no record was kept of the climatic conditions during the experimental period. It may also have been unwise of him to dig up the slab of stone at regular intervals, thus disturbing the toad's attempts at reaching a hibernating state, and exposing them to the light. Buckland's theory that insects had gained entry through a crack in the glass pane does not seem particularly likely. If the toad that had gained in weight had devoured such a considerable amount of insects, its cell would have been likely to contain toad excrements with the shell parts of these insects; if this had been the case, it is likely to have been noticed by an observer as astute as William Buckland. An alternative explanation of the animal's gain in weight may be that it was dehydrated when put in its cell, and later absorbed moisture through the skin.

The Blois Toad in the Hole

In June 1851, some French workers were digging a deep well near the city of Blois. They found a large flintstone, which was split in two by a hard blow with a pickax. In the middle of the stone sat a large toad. It appeared greatly astonished again to see the light of day. The toad in the hole jumped out and started to crawl away, but it was captured by the workmen, who carried it in triumph to the Society of Sciences in Blois. Both the toad and the stone were put in a damp cellar and embedded in moss. The French journalists had a field day speculating how many thousand years this toad had existed within its stone. The whole thing was termed "mysterious" and "absolutely inexplicable," and it seemed as if the Britons' morbid interest in the toad in the hole had spread, like some strange contagion, across the English Channel. The hammer that shattered the stony encasing of the toad in the hole would also shatter the fundaments of science.

The French Academy of Sciences was a more skeptical body, and they were dismayed to "see this ancient vulgar error once again raise its ugly head," as expressed by one of its more eloquent fellows. In view of the strong popular interest in the Blois toad, they appointed a committee of experts to look into the case. It was chaired by the veteran herpetologist Professor

A drawing of the Blois toad, as reproduced in *La Nature* of 1885.

André Duméril, who had taken part in the debate about the French rains of frogs and toads 17 years earlier. Although he was a convinced skeptic and was much inclined to disbelieve the toad-in-the-hole phenomenon altogether, he could not help being impressed by the way the toad's body exactly fitted the cavity in the flintstone. The members of the committee also noted that the toad's jaw had the mark of an irregularity in its hole, indicating that it had been growing within its lair for a prolonged period. After their investigation was concluded, Professor Duméril and the three other anatomists and zoologists in the deputation from the Academy of Sciences had to declare that they could find no evidence whatsoever of a fraud, and that the toad had apparently been situated within the flintstone, without any communication with the outside world. They confessed themselves completely baffled by this strange occurrence, and none of them even endeavored to explain it.

When Professor Duméril's paper was published in the Academy's *Comptes Rendus,* several other academians were aghast at this revival of an old fallacy, which had now been given support by one of their own number. They objected that André Duméril and his colleagues had been overly credulous, and that they had naively given the sanction of the academy to a ridiculous old error. Professor François Magendie, the great neurobiologist, suspected that the whole thing was an imposture, instigated, perhaps, by the workmen who had pretended to find the toad. He suggested that the toad should have been killed and dissected as soon as it was taken out of the stone; if incompletely digested insects were found within its intestinal canal, it was of course highly unlikely that it had spent many years inside the flintstone without food. But Magendie's (somewhat belated) advice was not acted on. At the next session of the academy, both Duméril and Magendie were thanked for their contributions to zoology, but the academy did not make any kind of official statement with regard to the Blois toad, probably owing to the persistent rumors that it was a clever hoax.

Scandal at the Exhibition

In the mid-1800s, the British nation's unhealthy fascination for the toad in the hole reached the height of a veritable mass hysteria. Every year, several novel cases were reported, both in the scientific periodicals and in the daily newspapers. The editor of the *Zoologist* magazine, Dr. Edward Newman, was deluged with letters and case reports, which he did not appreciate at all, since he was a firm opponent of the toad-in-the-hole phenomenon. Ironically, he wrote in one of his leaders that a considerable proportion of the toads of Britain were likely to spend all their lives within flowerpots, owing to his countrymen's abnormal interest in experimentation on toad longevity. In 1859, Charles Dickens read one of the articles in the *Zoologist,* which he later reviewed in his own journal *All the Year Round,* comparing the toad in the hole with the equally controversial phenomenon of toad showers: "not content with puzzling me with their subterraneous doings, these provoking reptiles are said to come down from the skies in showers." The British public, whose knowledge of natural history was otherwise unremarkable, knew at least two things about toads: they could fall from the clouds in showers, and they could live thousands of years embedded in solid stone.

In the autumn of 1862, just when Charles Dickens' article had been published, the great international exhibition was opened in London. In the eastern annex of the buildings were geological specimens from English and Welsh mines: one of them was a large block of coal from the mine of Cwm Tylery, which, when cloven in two, was found to contain a living frog. The newspaper writers all gathered round the frog's exhibition case, and the Londoners rejected the curiosities of three continents in favor of the Welsh frog in the coal, which was soon acknowledged as the leading light of the exhibition. In a sermon, a clergyman urged all Englishmen to watch this frog with reverence: it was one of the creatures first created by God, and had "breathed the same air as Noah and sported in the limpid streams in which Adam bathed his sturdy limbs."

Dr. Frank Buckland, the son of William Buckland and assistant surgeon to the Second Life Guards, shared his father's interest in the toad-in-the-hole problem. He was dismayed to see the renewed popular interest in this question, and to debunk the newspaper reports about this "antediluvian" frog, he obtained permission to examine it closely. It seemed to be a young frog rather than a fully grown specimen. The alleged century-long stay in the block of coal had not diminished its vivacity, and it jumped merrily about. Frank Buckland was seconded by his friend Professor Richard Owen, who suggested that frog spawn could have fallen into the pit and developed in some pool of water there. After having developed, the frog could have crawled into some dark crevice or fissure in the coal, to come tumbling out when the block of coal containing this fissure was broken by the workmen. It is a testimony to the British nation's great interest in the toad-in-the-hole controversy that the contributions of Buckland and Owen on this subject were not published in some obscure naturalist's journal, but as letters to the editor of the *Times*. In the *Punch* magazine, the frog was honored with a jocular poem:

Oh, who is this toad in a hole,
With face so expressively dark,
Who spends all his life in a coal,
And only comes out for a lark?

From Grub Street to Bridgewater Place
This Opéra comique*'s all the go;*
Where Buckland does alto and bass,
And Brown, Jones and Scroggins Buffo.

Several other rationalists wrote to the *Times*, to ridicule the ignorant populace for its persistent belief in the toad in the hole and other dubious phenomena, and to demand the instant expulsion of the frog and the block of coal from the exhibition. Some irate skeptics even suggested that the exhibitor of the frog be charged with fraud and imposture; the commissioners of the exhibition, particularly the liberal politician Lord Granville, were accused of being credulous fools. Meanwhile, members of the public defended the frog's honor in no uncertain terms, and more new cases of a toad in the hole were reported than ever before—three a month during the autumn and winter of 1862. Many of these were second- or even third-hand accounts and seem rather unconvincing; one newspaper article came from a man who had found a live toad inside his marble mantelpiece, when it has tipped from its position and broken in two. Richard Owen was sent many specimens of toad in the hole by mail or special delivery, but the busy academic turned them over to his wife Caroline for examination; she managed to expose at least one of them as a fraud, since the toad was much larger that the cavity in a block of coal it was said to have inhabited. Another living toad, which was found embedded in mortar when a wall was pulled down inside the church of north Wingfield, was sent by the vicar to Frank Buckland's home address. Buckland presumed that "the ecclesiastical toad" had been protected by divine powers during its perilous journey; having merely been put inside a paper parcel by the unworldly vicar, the animal, which was presumed to be several thousand years old, would have perished ignominiously if some careless railway porter had put a pianoforte on top of it! Otherwise, Buckland declared himself to be wholly unimpressed by this new specimen: it did not seem to be more than a year old, and leapt merrily about in its bowl.

Although the debate about the frog in the coal was a victory for Frank Buckland and Richard Owen, they failed in their objective to get the frog expelled from the exhibition. Nor did the British public seem to be wholly convinced by their arguments, although Mr. Punch mocked the Welsh frog in the coal:

Then what awe must each bosom overspread
As we gaze on that petrified bark;
On the bust of this quaint figure-head
That has yachted with Noah in the ark:

When we think that these somnolent eyes
With morning primæval awoke,—
That this solo (though sweet for its size)
Preluded Labyrinthidon's croak!

The Antediluvian Toads

Some commentators have marveled at the extreme interest in entombed toads in mid-nineteenth-century Britain: this obscure topic seems to have been given more attention in the newspaper press than any other biological problem. At least part of an explanation may be that the toad in the hole played a part in the debate about Darwinism. Certain conservative clergymen enrolled the entombed toads among their ranks to fight Charles Darwin's blasphemous theories. They considered the toad in the hole as one of the firmest arguments that a universal deluge really had happened. According to their version of geology, the rocks had later been formed from sedimented matter left by the deluge; certain toads, which had survived the Flood, had thus been encased in stone and lived there for thousands of years.

It was thus no coincidence that many of the British toads in the hole were described by clergymen. In April 1865, the debate was refueled after some workmen claimed to have found a live toad in a large block of magnesian limestone, at a depth of 25 feet from the surface of the earth. The cavity was no larger than its body and presented the appearance of being a cast around it. The toad's eyes shone with unusual brilliancy, and its mouth was found to be completely closed. The claws of the hind feet were unusually long. The Reverend Robert Taylor, Rector of St. Hilda's Church, Hartlepool, adopted the toad and gave it a home in the Hartlepool Museum. The rector reported to the press that the toad was still living several days later and seemed to be in excellent health, in spite of its inability to take nourishment; although its mouth remained closed, it occasionally uttered a "barking" noise through the nostrils. He claimed that the toad was 36 million years old, and that it had been sitting in its hole since the stone was first formed. Both the toad and the bold geologist parson became national celebrities. A party from the Manchester Geological Society visited Hartlepool to see the toad and inspect its hole in the block of stone. One of the geologists put his hand inside the hole and felt the marks of a chisel! He accused the collier who claimed to have found the toad of having perpetrated this fraud. The man first persisted that he was telling the truth, but when the geologist threatened to bring the stone and toad to the Natural History Museum in London for a closer examination, he refused and ignominiously withdrew. After this exposure, the rector also withdrew his daring statements about the antediluvian toad, but this did not prevent him from being much laughed at during one of the Manchester Geological Society's monthly meetings, where this revival of an old vulgar error was greeted with amused incredulity.

The Victorians' fascination with the toad in the hole slowly faded away during the 1870s and 1880s. The final outbreak of the hysteria occurred in 1901, when a certain Mr. W. J. Clarke, who was sitting in front of the fire in his snug Rugby home, suddenly gave a start when a large coal fell from the fireplace. He struck it with a poker, and it burst in two; a live toad fell out from its midst. In the fanciful newspaper accounts, this toad was stated to lack both a mouth and a rectum. Like some of its predecessors, the Rugby toad in the coal became quite a celebrity: it was even the subject of stereoscopic photographs available at the time.

During the remainder of the twentieth century, little has been written about the once-famous toad in the hole, in Britain or elsewhere, and the phenomenon has gradually become largely forgotten. Not even zoologists specializing in herpetology (amphibian biology) are aware of this obscure piece of toad lore, although some ethnologists have used the toad in the hole to exemplify the ridiculous nineteenth-century popular fallacies in natural history. But this is not the end of the insults directed toward the wretched subterranean toads by modern rationalists. When the once famous expression "toad in the hole" is looked up in the *Oxford Modern Dictionary,* the reader is informed that it can denote either a game or a dinner recipe, but nothing else. Both these attributes were derived from the legend of entombed toads, however. The objective of the game toad-in-the-hole, which dates from the seventeenth century, is to throw discs of lead into holes in a wooden structure; the piece of lead represents the toad, and the wood the stone surrounding its lair. According to competent ethnologists and local historians, games of toad-in-the-hole were played at some old-fashioned Sussex public houses as late as the 1930s. I have made inquiries in several public houses, in London as well as in Sussex and Kent, about the possibility of playing toad-in-the-hole, but none of the landlords have even seen it played, although some have vague memories of such an old-fashioned pub game. The only possibility of enjoying a game of toad-in-the-hole would be that some amusement arcade entrepreneur would construct a computerized version of this ancient pastime, to suit the apathetic "lager louts" of the present age.

It is even more remarkable that toads in the hole can actually be eaten. In the late 1700s, when the toad in the hole first gained notoriety, a dish made by seasoning a beefsteak and then baking it in batter, became highly à la mode. The beefsteak symbolized the toad, and the batter the surrounding stone. The dish survived for many years, but owing to austerity, in the 1920s and 1930s, the beefsteak was replaced by a brace of sausages. In 1934, T. S. Eliot wrote about "restaurants where you could get sausage and mashed or toad-in-the-'ole for twopence"; one wonders if he knew the true origin of the word! Not many London restaurants are prepared to serve up a toad in the hole today; anyone hoping to order such a dish from the Thickhead Pizza Co. or the Pukka-pukka Curry Inc. would have to go away disappointed. It is true that certain supermarkets sell what is called frozen toad in the hole, with foul-smelling grease and a disgusting brace of sausages instead of a beefsteak; this dismal dish could be recommended only to one's worst enemy. I have eaten a homemade toad in the hole made of beefsteak and batter, cooked by a devotee of traditional British cooking. A healthy appetite is needed to digest a dish of these contents, however; indeed, anyone might have lost his appetite when the "toad" was served in a steaming, huge pie dish. Three people and a Newfoundland dog received a hearty luncheon on this extraordinary meal.

Old Rip

On January 28, 1928, 2,000 people gathered round the derelict old courthouse of the little town of Eastland, situated near Fort Worth, Texas. One might have presumed that it was an election, a cattle auction, or a distribution of free beer, but it was an unprecedented incident in these parts that had roused most of the local population. When the old courthouse had been built 31 years earlier, a young local named Will Wood, who was known as a great practical joker, suggested that they were to put some things into the great hollow cornerstone of the building. The master builder put in a Bible and some newspapers and photographs, but Wood himself put a horned toad (*Phrynosoma cornutum*) into the stone before it was closed. He wanted to test the local legend that this lizard-like reptile, which occurs in Mexico and the southern states of the United States, could survive for more than 100 years without food, water, or air. There was much talk about Will Wood's experiment, and speculation was rife as to whether the horned toad was still alive. Several bets were wagered, and it was frequently suggested that the stone was to be taken out of its fundament, but Wood always refused. When the courthouse was to be pulled down, the moment of truth had arrived. Judge E. S. Pritchard was entrusted to remove the cornerstone, assisted by the three parsons of the community. Excitement grew when the stone was removed; the judge took out the Bible and the newspapers, and the Methodist pastor, the Reverend Frank E. Singleton, reached in and pulled out the blackened, lifeless body of the horned toad, holding it by one of its hind legs. Suddenly, the animal's other hind leg started twitching, and it violently gasped for breath. The pastor dropped the horned toad, which fell into the sand. There was a great uproar among the onlookers, and shouts of "The critter's alive!" and "Holy Jehosaphat!" The horned toad was carried in triumph to the town hall, where it was named "Old Rip" after Rip van Winkle, the fictional character who slept for 20 years; Old Rip had slept for 31!

The local newspapers all published Old Rip's picture on their front pages, and the Texas horned toad soon became a media celebrity. Reporters and tourists from all over the country visited Eastland to see its foremost attraction. Old Rip was exhibited in the window of a grocery store and slept there in his bowl during the night. One morning, Will Wood and his cronies were amazed to see that the famous reptile had vanished, as if swallowed by the ground. There were soon rumors that a local crook named C. F. Sheppard had kidnapped Old Rip to take him on tour. The toad-napper got as far as El Paso before being apprehended and returned to Eastland, where the judge placed the reptile in Will Wood's custody. A Dallas showman named Dick Penney sued Will Wood for more than 6,000 dollars, claiming to have a 10-week contract for the exhibition of Old Rip, but without success. Instead, Will Wood himself became Old Rip's manager, and exhibited the horned toad in various venues all over the United States. The horned toad was particularly successful in New York, and history has it that President Calvin Coolidge broke several other engagements to receive Old Rip in a personal viewing at the White House. When Old Rip returned to Eastland after his grand tour, he was the model for many posters and postcards. A scholar from the Texas Christian University, Dr. W. G. Hewatt, examined Old Rip and concluded that it was a perfectly normal specimen, whose horns about the head region were considerably worn, indicating that it was quite old. In January 1929, Old Rip was found dead in his bowl. No autopsy was performed, since Will Wood wanted to keep the body intact, but the cause of death was presumed to be pneumonia. Old Rip was embalmed in the funeral parlor of Mr. Ben Hamner, the Eastland mortician, and laid out in state, in a little blue coffin lined with purple velvet and white satin, which was put in a marble sarcophagus in the lobby of the Eastland County Courthouse.

Thus, even after his last croak, Old Rip remained the most famous citizen of Eastland. One morning in October 1930, however, something terrible happened: it was seen that the glass top of the marble tomb had been removed, and Old Rip's coffin was empty! The toad-nappers were soon hunted down, and the corpse of Eastland's most distinguished resident was returned to its resting place. He remained in peace until 1961, when he was kidnapped for the third time by some college students, who later shamefacedly returned the limp body in a paper bag, after having read the bloodcurdling curses and threats against themselves in the local newspapers. The year after, Old Rip entered politics. When one of the candidates for governor of Texas, Mr. John Connally, made a stop in Eastland during an electioneering tour, the local Democrats decided to give him their most valued possession. During previous stops on his tour, Mr. Connally had been given golden keys and giant pumpkins by the local authorities; in a ceremony in Eastland, the little blue casket containing Old Rip's limp body was handed over to the startled politician, who believed it to be a joke in rather bad taste. He picked up the embalmed body and swung it about by the leg, to the consternation of the old Eastlanders, who knew its age and fragility. Later, he "forgot" to return the toad's corpse when he left Eastland, and the president of the Eastland Chamber of Commerce had to drive to Cisco at breakneck speed to recover it. Rather shamefacedly, Mr. Connally returned Old Rip's body minus one hind leg, perhaps owing to further rough treatment by the candidate's entourage. Since 1962, Old Rip's body has been allowed to repose undisturbed in its marble tomb, admired by the Eastlanders and by thousands of tourists, many of whom have come merely to see the supernatural horned toad. Many zoologists have been disposed to doubt the tale of Old Rip, since horned toads rarely live for any prolonged period of time; it has also been hinted that the cunning Will Wood may well have substituted another horned toad for the corpse of the original one, just before the cornerstone was opened.

The Toad in the Hole on Trial

If the toad in the hole were to be summoned before judge and jury, charged with fraud, imposture, and breaking the laws of nature, its defense counsel would not be lacking in arguments to support the claim that his client was a reality. As exhibit one, he could present a file containing more than 210 cases of frogs and toads found inside stones, lumps of coal, or within the trunks of large trees. There have been reports of entombed toads from all over the world: from Europe, the United States, Canada, Africa, New Zealand, and the West Indies. The earliest of them are from the late fifteenth century, the latest occurred in Australia, in 1982. Many times, the frogs or toads were seen by several independent witnesses; not infrequently, as with Mr. Peters of Connecticut and the Swedish master builder, these people were entirely ignorant that similar prodigies had ever been described. After a close study of twenty of the best attested cases of entombed toads, the immediate conclusion is that the legend of the toad in the hole surely cannot have been based on imagination alone. Certain remarkable details about the aspect of the subterraneous toads and frogs often occur: the mouth was closed, and covered with a viscous membrane; the eyes shone brightly (perhaps owing to a maximal dilatation of the pupil, to adjust to a life in constant darkness); and the color of the toad's skin was darker than usual. The last phenomenon was also observed in some of the nineteenth-century experiments with entombed toads, which would support the notion that, at least in these cases, the toads had lived within their stony enclosure for a considerable period. Some early chroniclers of the toad in the hole claimed that the toads usually died when liberated from their prison; this happened in a number of instances, it is true, perhaps because the animal did not cope with the adaptation from its hibernating torpor to an active life. These cases are a minority, however, and several of the alleged entombed toads were stated to have been in excellent health and vigor immediately on being released, something which in itself arises suspicion.

The prosecutor in the trial of the toad in the hole would likely claim that the existence of living amphibians within solid stones is contrary to scientific opinion and common sense alike. Such a phenomenon is not only manifestly impossible: it is both pointless and irrational, and no reasonable explanation can be found as to why these toads preferred a living death to the active existence of the rest of their tribe. Twentieth-century zoologists have shown little interest in the toad in the hole; the few daring to comment on this "damned" phenomenon have sneered at this ridiculous old vulgar error in natural history, likening it to the popular fallacies that hedgehogs were milking the cows in the fields, or that living snakes could live for years as parasites in the human stomach. The herpetologist Professor G. K. Noble was one of the last to mention the toad in the hole, in his authoritative textbook *Biology of the Amphibia,* published in 1931. But he only bluntly declared that "the story that a toad can live for centuries entombed in stones or in old wells is sheer fable." The toad in the hole is not mentioned in any present-day herpetological textbook of repute. After being forgotten, or ignored, by the zoologists, the toad in the hole has fallen into bad company. It has been adopted by the modern pseudoscientists and devotees of paranormal phenomena, and vulgar books on "the unexplained" have put the subterranean toads on par with UFOs, ghosts, crop circles, spontaneous human combustion, and the horrors of the Bermuda Triangle. The most ridiculous hypotheses are employed to explain the entombed toad phenomenon: some authors still advocate the theory of a universal deluge, and others claim that the unsuspecting amphibians have been teleported into the rocks by mysterious powers. It may be asked why only frogs and toads should be teleported in this way, but one of the Fortean tomes contains a somewhat unconvincing discussion of the teleportation of termites, and it is true that there were some nineteenth-century instances of snakes and lizards found enclosed in cavities underground. Had the Forteans been better read in the annals of entombed animals, they could have reported some really juicy old stories. For example, Oliver Fiske, M.D., of Worcester can be credited for describing (in the memoirs of the American Academy of Arts and Sciences of 1803) the only known case of a "mouse in the hole." In another example, a farmer digging in a meadow found a hard, solid lump the size of a goose egg, which he broke with a spade; inside, he was astonished to see a little mouse, which "left a vacancy of the same dimensions as its size." After being warmed up in front of the fire, the little mouse "was restored to a perfect, living state, and ran off with activity"; it was speculated that the mouse had been immured for a century. An even more ludicrous tale can be found in the *Physica Curiosa* of Gaspar Schottus: in England, two dogs "of fierce countenance" were found enclosed in a block of stone from a quarry! They were as big as greyhounds but had no hair, and gave off an infernal smell. One died, but the other was taken to the Bishop of Winchester, who is recorded to have played with this malodorous, subterranean canine for several days.

If the temporal distribution of the more than 210 cases of toad in the hole is studied closer, it is seen that these observations of entombed toads became increasingly frequent from the early seventeenth century onward. M. Guéttard, writing in the 1780s, was quite right to consider this phenomenon one of the great unsolved biological riddles of his time. The remarkable series of toad-in-the-hole observations culminated in the mid-1800s, but there were a steady trickle of new cases well into the early twentieth century. In our own time, far fewer subterranean toads have been discovered, and the rationalists have of course used this as an argument that the entombed animals of olden times belong to the realm of mythology rather than of zoology. The friends of the toad in the hole object to this reasoning, stating that the use of modern mining technology has been disastrous to the subterranean amphibians: instead of being carefully freed from its hole by the quarryman, the entombed toad is ground into mincemeat by the powerful mining drills. Another factor likely to decrease the number of entombed toad cases is that the people finding them are afraid of being ridiculed when reporting such an odd and "impossible" occurrence. The time when the toad in the hole was debated by learned academies is long past, and the report of an entombed toad would today be discarded, with scorn, by even the funny pages of the tabloid newspapers. When an Australian zoologist made a broadcast about entombed toads in 1983, however, treating the subject quite seriously, several people contacted him to describe their own experiences of toads and frogs in stone.

The geographical distribution of the toad in the hole phenomenon is also of interest. Not fewer than 150 of the 218 cases known to me are from the United Kingdom; 105 of these were described in the nineteenth century, many of them during the 1850s and 1860s, when interest in this phenomenon was at its peak. Although the toad in the hole lore was well known all over the civilized world at the time, one marvels at the strong British interest in this obscure question. The religious debate about Darwinism contra the Deluge, into which the entombed toads were recruited by certain conservative clergymen, may have played its part, but it cannot explain all aspects of the great popular interest in this phenomenon. It may be that the toad, pent up in a block of granite and one day freed through a mighty blow from the midwife stonecutter's hammer, had a strange symbolic significance for the repressed Victorians. For some individuals, it seems to have become a veritable *idée fixe* that the toad in the hole really existed, and they even went as far as to forge preparations with dead or living toads encased in blocks of stone or coal. The Reverend Taylor's toad is likely to have been a hoax, as well as at least one of those sent to Professor Owen; there were also surreptitious rumors that the famous frog in the coal at the exhibition was a forgery, prepared to advertise the mine where it was said to have been found. During the heyday of entombed toad lore, the two zoologists Edward Newman and John Plant traveled to several mines and quarries and chatted with the workmen, buying them rounds of ale and porter in the pubs. They promised that the miner who showed them an undoubted toad in the hole would be richly rewarded. The miners and quarrymen eagerly drank their ale, bragging that, since entombed toads were frequently occurring in the rocks they were working, they would soon be claiming their reward. After several years, no toad in the hole was forthcoming, although Plant visited the mines and quarries several times. The workmen often said that one of their colleagues, who had recently moved to another mine, had once discovered a toad in the hole, but Plant remained incredulous.

A Gloucester naturalist, who promised a guinea to the quarryman who could show him a toad in the hole, received a specimen soon after. But the presumedly antediluvian toad died the same evening! When dissected, it proved to have several half-digested insects in its stomach, a diet that it could not very well have attained while pent up in a block of granite for thousands of years. In the 1860s, the Victorians' unwholesome interest in self-immured amphibians had risen to such a peak that there was even a market for the mass production of faked toads in the hole. In his workshop in Leeds, an individual called Toad Jack baked toads in an oven to make them black, before putting them into bisected pieces of rock or lumps of coal, in which a hole corresponding to the animal's body had been carefully cut out. These forged specimens were readily purchased by avid curiosity seekers; not a few of these souvenirs were bought by publicans, who exhibited them in their ale houses. A toad in the hole attracted more visitors than a comedian or pub musician, and caused the ale barrels to be emptied faster than either of these attractions.

A closer study of the many English nineteenth-century toad-in-the-hole cases shows that many of them were reported in the newspaper press without receiving notice in any scholarly journal. Many, but by no means all, of these still seem quite trustworthy. Some of the newspaper accounts about "A Wonderful Toad" or "Remarkable Fact in Natural History" have almost exactly the same wording. It does not seem overly fanciful to suspect that certain unscrupulous hack journalists copied interesting articles from old newspapers and sold them to their editors as fresh news; this practice is known to have been widespread in the mid-1800s, and may not be totally extinct today. Articles about entombed toads were particularly attractive, owing to the great topical interest of this phenomenon, which was known by all and sundry. Some other newspaper stories about toads in the hole being discovered are likely to have been hoaxes, perpetrated by clever pranksters who wanted to make fun of the gullible, sensation-seeking journalists.

The result of the most audacious of these practical jokes was published in the *Illustrated London News* of 1856. When a railway tunnel was being blasted through a mountain between Saint-Didier and Nancy, the workmen were horrified to see a monstrous creature flutter out from a hole in a large boulder, which had just been cloven in two by an explosion. The creature was the size of a goose, with an abominably ugly head and a long beak armed with rows of sharp teeth. It uttered a hoarse cry and then fell down dead. A prominent French naturalist, it was stated, identified it as a long-extinct saurian flying lizard, a *Pterodactylus anas,* a species that had flourished during the Jurassic era. The clue to this hoax is that *anas* is the Latin word for duck: the erudite prankster must have had a jolly time when he saw the current issue of the *Illustrated London News,* and he must have exulted at having served the ignorant Victorian newspaper men such a juicy canard. It is surprising that this ludicrous "pterodactyl in the hole" was taken quite seriously both in

History of the Supernatural, by William Howitt, and in *Phenomena,* by J. Michell and R. J. M. Rickard. The former work was published in 1865—and the latter in 1983!

Richard Owen and Frank Buckland were probably right when they claimed that many of the nineteenth-century toads in the hole had not spent any prolonged period inside the stone. At the height of the toad-in-the-hole debate, it was sometimes enough that a toad or frog leapt forth just as a block of stone was being split for the entombed toad hysteria to suffer another outbreak. Frank Buckland suggested that many of the other cases might well have a perfectly natural explanation: an adventurous young toad might crawl through some rocky crevice to reach a natural hollow within a block of stone. It could hibernate in here with ease, living off the insects that ventured into the same crevice, as well as its own deposit of fat. In the early spring, the toad faced an unpleasant surprise: either it had grown in size, becoming too large to get out, or the hole had been plugged by gravel or calcific deposits; in any case, the toad was a prisoner within its block of stone, which was likely to become its coffin, if it was not freed by a quarryman to become a national hero in a flurry of newspaper publicity.

In zoological gardens, it has been noted that toads may live to be 35 years old if allowed to lead a sedentary life with ample provisions of food. The Victorian naturalist Thomas Pennant recorded that he kept a toad as a family pet for 36 years. Each evening, this venerable toad emerged from its lair underneath the steps of his house and politely waited to be picked up and put on the dining table, where it was given a meal. The question of how long a toad can live in a hibernating state, enclosed within a stony cell, is still unsolved and is likely to remain so. Some of the older experiments speak in favor of the tenacious amphibians being able to live for many years in a hibernating state without access to food. If some demented Ph.D. student would today attempt to reproduce William Buckland's experiment, he or she would run into considerable difficulties with both the university board and the Royal Society for the Prevention of Cruelty to Animals. Since the Ph.D. student's grant is limited to 3 years, he would have to rely on the wretched toad's expiring within this time; if the toads were still living after 3 years, he would himself have to live on thin air just like his experimental animals, if he had not previously been torn to pieces by members of the militant animal rights movement.

During the heyday of the toad in the hole, many of the specimens were recorded to have been presented to various museums, but the vast majority of these relics of the credulous Victorian era seem to have been discarded by

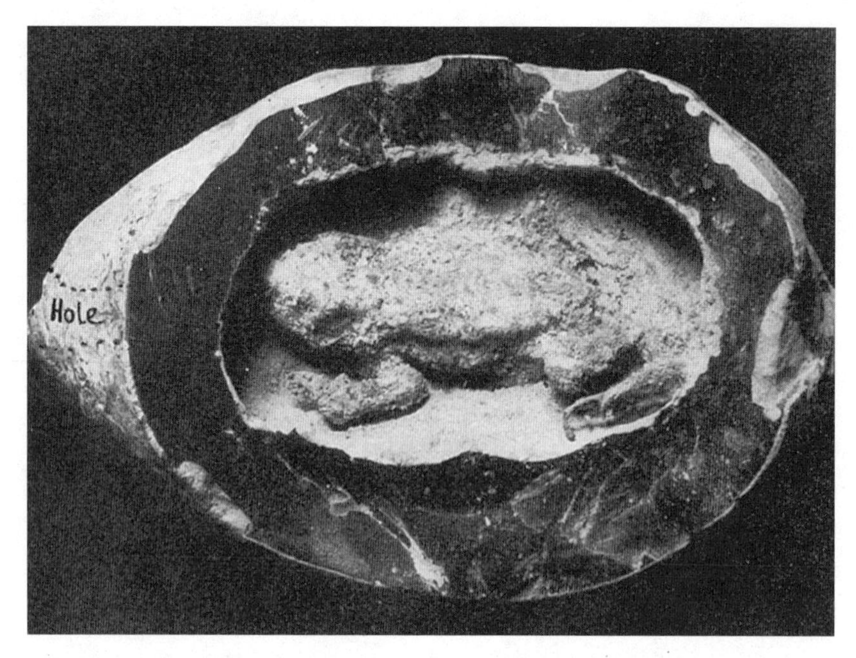

A contemporary photograph, published in the *Illustrated London News*, of the flintstone nodule containing a dead toad discovered by Charles Dawson. Reproduced with permission of the British Library.

the latter-day museum curators. It is recorded in 1821, for example, that Lord and Lady Duncan donated a toad in the hole to the Edinburgh University Museum, which is now part of the Royal Scottish Museums; the toad is no longer within its collections, however. Nor has a lump of rock, containing a live frog found at a depth of many feet below ground, which was exhibited at the museum of the Midlands Institute in Birmingham in the mid-1800s, been kept for posterity. The world's only toad in the hole today resides in Brighton, in the Booth Museum of Natural History. It was donated by a certain Mr. Charles Dawson, a lawyer and amateur naturalist, in 1901. The preparation consists of an oval, hollow flintstone nodule, containing a long-dead, desiccated toad. Dawson said that he had been given it by some workmen, who had noticed that the stone seemed lighter than could be expected and broke it with a spade to see what was inside. The flintstone, which is probably 75 to 100 million years old, is likely to have been formed around one of the primitive sponges extant during this period. These sponges were formed like a bag, with a narrow, spoutlike stem. At a later period, water must have entered the stone and dissolved the sponge. There was a connection between the hole inside the stone nodule and the outside world, by way of the narrow channel once occupied by the sponge's stem; this connection can still be seen in the Brighton preparation. It may well be that a young toadlet crawled in through this aperture. In some way, it was able to procure food in there; perhaps some insects and larvae chose the same way in. Finally, the toad grew in size, to the point of being unable to get through the narrow passage through which it had entered; after some time, it starved to death within its stony prison. If the workmen had split the stone before the animal had died, it would have joined the ranks of the traditional toads in the hole. The Brighton stone nodule is unlikely to have been the only one formed around a sponge, and the explanation detailed here can be extended to cover quite a few of the other toads in the hole—and perhaps even provide a tentative explanation of the phenomenon. There is, however, one serious objection to this theory. The name of the amateur naturalist Charles Dawson,

The Brighton toad in the hole today, reproduced with permission of The Booth Museum of Natural History, Brighton U.K.

who donated the toad to the museum, was blackened by his involvement in the most notorious scientific forgery of the twentieth century: the Piltdown scandal, in which a faked anthropoid skull was claimed to belong to Darwin's "missing link" between human and ape. Several anthropologists have even accused Dawson of having forged the skull and other Piltdown "fossils," including a piece of elephant's bone carved to look like a cricket bat—a fitting accouterment to "the first Englishman"!—with his own hands. Although some later evidence instead points toward the London zoologist Martin Hinton as the main perpetrator of the hoax, Dawson is still one of the suspects, and his other "discoveries" must be viewed with skepticism. Comparison with the original 1901 photographs demonstrates that the Brighton toad has shrunk much in size since then, indicating that it was quite recently dead and dried up when submitted to the museum. Another suspicious circumstance is that it succumbed neither to mold nor to fungal attack, but to mummification.

The toad in the hole is an anomaly in the true sense of the word. The phenomenon is not only irrational but completely inexplicable, and there is no reasonable explanation to this remarkable series of observations of toads and frogs discovered inside blocks of stone. The counterarguments proposed here are hardly enough to completely exorcise the unblessed, prematurely buried toads, which have led their own slumbering, accursed half-life outside the boundaries of biology for several centuries. It is a fitting final tribute to the entombed toads to quote from Dante Gabriel Rossetti's poem *Jenny,* written in 1870:

Like a toad within a stone
Seated while time crumbles on;
Which sits there since the earth was cursed
For man's transgression at the first;
Which, living through all centuries,
Not once has seen the sun arise;
Whose life, to its cold circle charmed,
The earth's whole summers have not warmed;
Which always—whitherso the stone
Be flung—sits there, deaf, blind, alone. . . .

Sources

The Dancing Horse

The best two essays on Mr. Banks and his horse are those by J. O. Halliwell-Phillipps, *Memoranda on Love's Labour's Lost* (London 1879, 21–57, 70–73) and A. Freeman, *Elizabethan Eccentrics* (New York 1989, 123–139). Other worthwhile articles are those by C. E. Browne (*Notes and Queries* 5th Ser. 6 [1876], 387), S. H. Atkins (*Notes and Queries* 167 [1934], 39–44), and W. Schrickx (*Notes and Queries* 227 [1982], 137–138). Marocco is also mentioned in *The Old Showmen and the Old London Fairs,* by T. Frost (London 1881). The pamphlet *Maroccus Extaticus* was reprinted by the Percy Society (Vol. 9 [37], London 1844) and is discussed further in the Collecteana Anglo-Poetica (*Historical & Literary Remains* 52 [1860], 152–156). Modern performing and "clever" horses have been discussed by H. Cazier-Charpentier (*Le Cirque dans l'Univers* 122 [1981], 9–11), P. R. Levy (*Les Animaux du Cirque;* Paris 1991, 13–24), and D. K. Sandland (*Feral Children and Clever Animals;* Oxford 1993).

Lament of the Learned Pig

The two most important sources on learned pigs are the scholarly and very amusing book *Learned Pigs and Fireproof Women* (London 1986, 8–27), by the American magician and circus historian Ricky Jay, and the article by Dr. G. E. Bentley Jr. in the *Colby Library Quarterly* (18 [1982], 87–104). There are many prints, handbills, and newspaper cuttings about learned pigs in Lysons' *Collectanea,* in Miss Banks's Collection of Broadsides, and in the Menageries section of the Fillinham Collection of Cuttings from Newspapers, all three of which are kept in the British Library. *The Symbolic Pig,* by F. C. Sillar and R. M. Meyler (Edinburgh 1961, 59–67), and the articles in the *New Penny Magazine* (2 [1899], 768) and by H. Stepstone (*Royal Magazine* 8 [1902], 442–444) give further information about pigs in show business. The books *Tiere als Artisten* by A. Lehmann (Wittenberg 1956), *Les Dompteurs* by H. Thétard (Paris 1928), and *Les Animaux du Cirque* by P. R. Levy (Paris 1991) provide important information about trained animals in general. A background in eighteenth-century popular zoology and the anthropomorphizing of animals is given by K. Thomas in *Man and the Natural World* (New York 1983) and the articles by H. Ritvo (*Harvard Library Bulletin* 33 [1985], 239–279) and O. D. Oerlemans (*Mosaic* 46 [1994], 174–190). The books by Pinchbeck and Haney, quoted in the text, provide information about nineteenth-century animal training methods, which have been further discussed by Martin Gardner (*Scientific American* 240(5) [1979], 20–24).

310 The Feejee Mermaid

The mythology and physiology of mermaids is reviewed by, among others, P. Lum (*Fabulous Beasts;* New York 1951, 130–144), R. Carrington (*Mermaids and Mastodons;* London 1959, 3–20), G. K. Brøndsted (*Havfruens Saga;* Copenhagen 1965), and B. Phillpotts (*Mermaids;* London 1980) and in articles by A. S. Goodman (*Journal of Popular Culture* 17 [1983], 32–48), C. B. Fleming (*Smithsonian* June 1983, 86–95), K. Banse (*Limnology and Oceanography* 35 [1990], 148–153), and T. W. Pietsch (*Archives of Natural History* 18 [1991], 1–25). The most important sources on Captain Eades and his mermaid are the British Library's collections of pamphlets and handbills (Sadlers Wells; Morley's Scrap-book), and the valuable Lysons' *Collectanea,* as well as the 1822 and 1823 volumes of the *Gentleman's Magazine,* the *Annual Register,* and the *Times* and *Mirror* newspapers. William Clift's description of the mermaid is kept in the archives of the Royal College of Surgeons of England (Cabinet II.7), and the same file also contains copies of Home's correspondence with Eades and Clift about the mermaid. Another account of the mermaid, also by William Clift, with a drawing of it by the artist Robert Hills, is in the Natural History Museum, London (OC 62.8/39). Later articles about the Feejee Mermaid include those by L. E. Chaney (*The Dickensian* 50 [1984], 39–40), S. C. Levi (*Western Folklore* 36 [1977], 149–154), and H. Ritvo (*Victorian Literature and Culture* 19 [1991], 277–291). P. T. Barnum tells the story of the Feejee Mermaid in his *Struggles and Triumphs* (Hartford 1869), and his later biographers all have discussed the mermaid's career: see M. R. Werner (*Barnum;* New York 1923, 56–63), I. Wallace (*The Life & Times of P. T. Barnum;* New York 1959, 73–77), and N. Harris (*Humbug: The Art of P. T. Barnum;* Boston 1973, 62–67). Of particular value are the well-written and scholarly biography of Barnum by Mr. A. H. Saxon (*P. T. Barnum, the Legend and the Man;* New York 1989, 119–123) and the same writer's *Barnumiana* (Jumbo's Press, Fairfield CT, 1995, 38–39, 62–63), which add much new information about the mermaid. The 1822 mermaid pamphlet is at the Beinecke Rare Book Library at Yale University, and the 1842 pamphlet is at the Library Company of Philadelphia; neither is held by the British Library. Much valuable information about the Feejee Mermaid and other faked mermaid specimens has been obtained from Mr. A. H. Saxon, Ph.D., Bridgeport. Also thanked for valuable information are the archives of Harvard Peabody Museum (L. Wolf Whitehead, who kindly supplied valuable copies from the mermaid file), the British Museum (Prints and Drawings), the Museum of Mankind (Dr. Brian Durrans and Mr. Jim Hemall), and the Brighton Royal Pavilion Museum (Mr. Antony Shelton). Carl Linnaeus's writings about mermaids are in his *Bref och Skrifvelser,* Vol. 2 (Stockholm 1908, 129–131). Other faked mermaids have been described by Frank Buckland in volume 2 of his *Curiosities of Natural History* (London 1865, 113–122), by Henry Lee (*Sea Fables Explained;* London 1882, 213–217), by J. Boullet (*Aesculape* 41 [1958], 3–62), by J. Hutchins (*Discovering Mermaids and Sea Monsters;* Tring 1968), and in an anonymous article in the *Buried Treasures of the Peabody Museum* (2 [1969], 1–4). Another valuable source is the excellent book *Animal Fakes and Frauds* by Peter Dance (Maidenhead 1976, 37–56).

Obituary for an Elephant

Some excellent books on the natural and cultural history of elephants are *Elephants* by R. Carrington (London 1958), *Elephants Ancient and Modern* by F. C. Sillar and R. M. Meyler (Ontario 1968), *Elephants* edited by S. K. Eltringham (Poole 1991), and *Elefanten in Zoo und Circus* edited by A. Haufellner et al. (London 1993). Additional information can be found in the articles by N. Kourist (*Zoologische Beiträge* 18 [1971], 141–148) and R. Delort (*L'Histoire* 20 [1980], 32–40). The most important sources on Chunee and the Exeter Change menagerie are the collections of press cuttings, pamphlets, and drawings in the Enthoven Collection, Theatre Museum, London, and the Archives of Westminster City Libraries, London. Much additional material, including several scarce pamphlets and prints, is in the British Library (Bartholomew Fair, Sadlers Wells, and Lysons collections). Four important contemporary sources are E. Cross, *Companion to the Royal Menagerie, Exeter Change* (London, 1820); J. Taylor, *The Life, Death and Dissection of the Largest Elephant Ever Known in this Country* (London 1826); P. Egan, *Anecdotes of the Turf, the Chase, the Ring, and the Stage* (London 1827, 281–293); and W. Hone, *Every-Day Book*, Vol. 2 (London 1834, 321–345). The *Times*, *Morning Chronicle*, and *Mirror* newspapers of 1826 contain many articles about Chunee's tragic death and the elephant mania that ensued. Important information about Chunee's skeleton is contained in the manuscript "Diaries of William Clift" (entries for March 1, 1826; May 26–June 10, 1831; and Nov. 1, 1831), kept in the archives of the Royal College of Surgeons of England, and in volume 1 of the descriptive catalogue of the osteological series in the Royal College of Surgeons' Museum. Charles Dickens wrote about Chunee in *All the Year Round* (13 [1865], 256–257); two later articles were in *Once a Week* (Nov 1863, 586–588) and the *Picture Magazine* (April 1894, 205–206). The essential modern description of the Exeter Change and its animals is that by Professor Richard D. Altick, in his masterly *The Shows of London* (London 1978, 302–316); other accounts are C. H. Keeling's *Where the Lion Trod* (Clam Publications, Shalford [Surrey UK] 1984, 2–25) and *Where the Elephant Walked* (Clam Publications, Shalford [Surrey UK] 1991, 4–9, 41–43) and H. Ritvo's *The Animal Estate* (London 1990, 205–242). Mr. C. H. Keeling, Shalford, and Mr. Ian Lyle, Librarian to the Royal College of Surgeons, London, are thanked for valuable help.

Jumbo, King of Elephants

The only full-length biography of the King of Elephants is W. P. Jolly's *Jumbo* (London 1976). In its preface, it is remarked that the biographer of an elephant has the great advantage not to be overburdened with the personal papers of the deceased; nor is there any worry of offending the family! *London Zoo from Old Photographs 1852–1914* by Mr. John Edwards (London 1996) is another valuable source, and contains an annotated collection of almost every known photograph of Jumbo at London Zoo. Four important articles about Jumbo are those by J. Bannerman (*Maclean's Magazine* 68 [1955], 28–29, 43–50, 84), J. L. Haley (*American Heritage* 24 [1973], 62–68, 82–85), T. James (*Smith-*

 sonian 13 [1982], 134–152), and D. J. Preston (*Natural History* 92 [1983], 80–83). The article by S. L. Shoshani et al. (*Elephant* 2 [1986], 86–122) is up to date and has a particularly useful list of references. The biographers of P. T. Barnum, foremost among them being Mr. A. H. Saxon (*P. T. Barnum, the Legend and the Man;* New York 1989, 291–302), add a good deal of information about Barnum's purchase of Jumbo and the elephant's sojourn at his circus. In addition, J. R. Russell (*University of Rochester Library Bulletin* 3 [1947], 12–20), R. L. Carpenter (*Tuftonian* [1941], 6–11), and G. G. Goodwin (*Natural History* 61 [1952], 16–21, 45–46) have described the preparation of Jumbo's skeleton and hide. B. C. Landauer (*New York Historical Society Quarterly Bulletin* 18 [1934], 43–52) has studied the advertisement campaigns inspired by the famous elephant. Mr. A. H. Saxon, Ph.D., Bridgeport, and Mr. John Edwards, London, are thanked for valuable information.

Animals on Trial

Several books have been written on this subject; the earliest, and best, of these is *The Criminal Prosecution and Capital Punishment of Animals* by E. P. Evans (London 1906). Other works include *Tierstrafe, Tierbannung und Rechtsrituelle Tiertötung im Mittelalter* by H. A. Berkenhoff (Strasbourg 1937), *Les Procès d'Animaux* by M. Rousseau (Paris 1964), and *Les Procès d'Animaux du Moyen Age à Nos Jours* by J. Vartier (Paris 1970). Three important articles are those by K. Seifart (*Zeitschrift für Deutsche Kulturgeschichte;* 1856, 424–432), K. von Amira (*Mitteilungen des Instituts für Oestereichische Geschichtsforschung* 12 [1891], 529–601), and W. W. Hyde (*University of Pennsylvania Law Review* 64 [1916], 696–730). Later papers include those by J. J. Finkelstein (*Transactions of the American Philosophical Society* 71(2) [1981]), J. Heritier (*Histoire* 84 [1985], 72–76), L. Price (*Police Review* August 8, 1986), and P. Mason (*Social Science Information* 27 [1988], 265–273). Two particularly valuable recent articles are those by E. Cohen (*Past and Present* 110 [1986], 6–37) and M. Pastoureau (Histoire 172 [1993], 16–23).

The Riddle of the Basilisk

Many of the older sources are mentioned in the text. More recent articles on basilisks include those by H. Phillips Jr. (*Penn Monthly* 14 [1882], 33–42), A. Garboe (*Teologisk Tidsskrift* 3R. 8 [1917], 17–30), G. Polivka (*Zeitschrift des Vereins für Volkskunde* 28 [1918], 41–56), L. A. Breiner (*Isis* 70 [1979], 30–47), E. V. Walter (*Comparative Civilization Review* 10 [1983], 51–72), and R. N. Doetsch (*Pacific Discovery* 36 [1983], 25–30). The books *Die Zauberkraft des Auges und das Berufen* by S. Seligmann (Hamburg 1922, 183–195, 522–525) and *Animal Lore in English Literature* by P. A. Robin (London 1932, 84–95, 181–188) provide additional information. Faked basilisks have been described by A. Forti (*Atti del R. Istituto Veneto di Scienze, Lettere et Arti* 88 [1928], 225–238), E. W. Gudger (*Scientific Monthly* 38 [1934], 511–523), and F. Grondona (*Physis* 11 [1969], 249–266). Four excellent Danish sources are the article by H. F. Feilberg (*Naturen og Mennesket* 12 [1894], 164–196), and the books *Fabeldyr og andre Fabelvæsener* by J. Anker and S. Dahl (Copenhagen 1938,

192–197), *Thomas Bartholin*, Vol. 1, by A. Garboe (Copenhagen 1950, 110–112, 196–197), and *Fabeldyr og Sagnfolk* by B. Holbek and I. Piø (Copenhagen 1967, 341–353). Masculinized hens laying eggs have been discussed by, among others, L. J. Cole (*Journal of Heredity* 18 [1927], 97–106) and T. R. Forbes (*Yale Journal of Biology and Medicine* 19 [1947], 955–970). Strange eggs with "young basilisks" inside them have been described by C. Davaine (*Comptes Rendus des Séances et Mémoires de la Société de Biologie* 3. Sér. T. 2 [1860], 183–266), M. Henry (*Bulletin de l'Académie Vétérinaire de France* 1 [1928], 157–158), G. O. Hall (*Poultry Science* 24 [1945], 496–498), and in a valuable review by Bob Skinner (*Fortean Times* 46 [1986], 46–49).

Spontaneous Generation

The German book *Urzeugung und Lebenskraft*, by Professor Ernst von Lippmann (Berlin 1933), provides a broad overview of the older history of spontaneous generation. *La Genèse de la Vie* by J. Rostand (Paris 1943) and the *History of Magic and Experimental Sciences* by L. Thorndike (New York 1941) provide further information. Later articles include those by S. Lindroth (*Lychnos* [1939], 159–192), W. Capelle (*Rheinisches Museum für Philologie* NF 98 [1955], 150–180), D. M. Balme (*Phronesis* 7 [1962], 91–104), E. I. Mendelssohn (*Actes XIIe Congrès d'Histoire des Sciences* 1 [1971], 201–229), and R. Kruk (*Journal of Semitic Studies* 35 [1990], 265–282). In particular, the excellent article by D. Passmann (*British Journal for Eighteenth-Century Studies* 11 [1988], 1–17) provides a valuable overview of literary aspects of spontaneous generation.

The Boramez articles of Sir Hans Sloane and Dr. Breyn were published in the *Philosophical Transactions of the Royal Society of London* (20 [1695], 461–462 and 33 [1725], 353–360). *The Vegetable Lamb of Tartary* by Henry Lee (London 1887) is a standard work. Later articles include those by G. Schlegel (*Actes du 8:e Congrès des Orientalistes* 4 [1889], 17–32), B. Laufer (*Journal of American Folk-lore* 28 [1915], 103–128), A. W. *Exell (Natural History Magazine* 3 [1932], 194–200), A. F. Tryon (*Missouri Botanical Garden Bulletin* 43 [1955], 25–28), and I. Malaxecheverria (*Romance Philology* 35 [1981], 266–268). Sir Richard Lea's account is in the archives of the Royal Society of London (CC.P. XV[I].14). Mr. Michael Webb, of the Bodleian Library, and Dr. Arthur MacGregor, of the Ashmolean Museum, Oxford, are thanked for valuable advice concerning Sir Richard Lea's lambskin coat and its vicissitudes in Oxford. Mrs. John Nicholson, MBE, of the Museum of Garden History, London, is thanked for a detailed showing of the vegetable lamb there.

The most important source on the barnacle geese is the book *Barnacles in Nature and Myth* by E. Heron-Allen (Oxford 1928). M. Guéttard's article was published in volume 4 of his *Mémoires sur Différentes Parties des Sciences et Arts* (Paris 1783, 238–303). Later works worth consulting are the *Diversions of a Naturalist* by Sir Ray Lankester (London 1915, 108–141), and *The Folklore of Birds* by E. A. Armstrong (Cambridge 1959, 225–237), and the articles by F. Moll (*Archiv für die Geschichte der Mathematik, der Naturwissenschaften und der Technik* NF 11 [1928–1929], 123–149), J. Seide (*Centaurus* 7 [1961], 207–212), and F. Egmond and P. Mason (*Journal of the History of Collections* 7 [1995], 25–43).

The Spontaneous Generation Controversy from Descartes to Oparin by Dr. J. Farley (Baltimore 1977) is a standard work with regard to the eighteenth- and nineteenth-century debate on spontaneous generation. The articles by P. Smit (*Nieuwe Nederlandse Bijdragen tot de Geschiedenis der Geneeskunde en der Naturwetenschappen* No. 8 [1982], 169–185) and E. G. Ruestow (*Journal of the History of Biology* 17 [1984], 225–248) describe Antonie van Leeuwenhoek's campaign against spontaneous generation. The eighteenth-century debate is further discussed by S. A. Roe (*Isis* 74 [1983], 159–184) and M. Carozzi (*Gesnerus* 42 [1985], 265–288).

The modern debate on spontaneous generation and the origin of life has been described by K. Dose (*Interdisciplinary Science Reviews* 13 [1988], 348–356), L. E. Orgel (*Scientific American* 271 [1994], 53–61), and J. Maddox (*Nature* 367 [1994], 409). The fascinating recent book *Vital Dust* by Professor Christian de Duve (New York 1995) contains a broad overview of the history of life on earth. Some recent articles on the origin of life are those by G. von Kiedrowski (*Nature* 381 [1996], 20–21), J. P. Ferris et al. (*Nature* 381 [1996], 59–61), P. Parsons (*Nature* 383 [1996], 221–222), J. M. Hayes (*Nature* 384 [1996], 21–22), S. J. Mojzsis et al. (*Nature* 384 [1996], 55–59), and M. H. Engel and S. A. Macko (*Nature* 389 [1997], 265–268).

Odd Showers

Many older sources are referred to in the text. Carl Linnaeus's paper about the lemmings was published in Swedish, in the *Kungliga Vetenskapsakademins Handlingar* (1 [1740], 320–325). The French nineteenth-century debate about showers of frogs is reviewed by A.-M.-C. Duméril and G. Bibron in volume 8 of their *Erpétologie Générale* (Paris 1841, 223–233). A valuable older review of all kinds of odd showers is that by W. L. McAtee (*Monthly Weather Review* 45 [1917], 217–224). Dr. E. W. Gudger collected tales of numerous fish showers in three well-researched articles in *Natural History* (21 [1921], 607–619), *Annals and Magazine of Natural History* (10. Ser. 3 [1929], 1–26), and *Scientific Monthly* (29 [1929], 523–527). Two later brief reviews are those by R. E. Martin (*Popular Science Monthly,* July 1932, 24–25, 108) and M. W. Martin (*Science Digest* 67 [1970], 32–36). Herved Bernin wrote on rains of worms in an obscure pamphlet published in Swedish only: *Maskregnsproblemet och dess Vetenskapliga Lösning* (Lund 1925). Charles Fort's *Complete Books,* containing an immense collection of odd and macabre tales, were reissued by Dover Publications, New York, in 1974. The modern Forteans have written at length on this phenomenon; the books by S. Welfare and J. Fairley (*Arthur C. Clarke's Mysterious World;* London 1980) and by J. Michell and R. J. M. Rickard (*Phenomena;* London 1977; and *Living Wonders;* London 1982) both contain reviews of all kinds of odd showers. *Living Wonders* is fascinating to read and particularly well researched, but it lacks footnotes and references, and its conclusions are sometimes not to be taken seriously. Few modern meteorologists have written about these odd showers, but the papers by M. W. Rowe (*Journal of Meteorology UK* 7 [1982], 177) and D. Elsom (*New Scientist* 118 (1615) [1988], 38–40) provide much valuable information. An excellent, annotated review of the literature is contained in the book *Tornados, Dark Days, Anomalous Precipitation, and Related Weather Phenomena* by Dr. W. R. Corliss (Glen Arm 1983, 39–80).

Toad in the Hole

The tale of Mr. Gråberg and his stone frog was originally published in the *Kungliga Vetenskapsakademins Handlingar* (2 [1741], 248–255). Dr. Pihl's discussion is kept in the archives of the Swedish Academy of Sciences (sekr. arkiv 1), and A. Ph. Tidström's paper is kept in archives of the University Library of Uppsala (D 1457, f. 161). The Swedish stone frog is further discussed by E. W. Dahlgren in volume 1 of the *Svenska Vetenskapsakademins Protokoll* (Stockholm 1918, 340–357) and by C. Sahlin in the Swedish mining periodical *Med Hammare och Fackla* (1938, 128–136). Five important early articles on subterranean amphibians are those by C. N. Le Cat, in Alléon Dulac's *Mélanges d'Histoire Naturelle* (3 [1764], 95–105), J. S. Guéttard (*Mémoires sur Differentes Parties des Sciences et Arts;* 4 [1783], 615–636, 683–684), W. Buckland (*Zoological Magazine* 5 [1831], 314–320), M. Vallot (*Bibliothèque Universale des Sciences* 56 [1834], 251–266), and A.-M.-C. Duméril (*Comptes Rendus des Séances de l'Academie des Sciences* 33 [1851], 105–116). Gilbert White's manuscript account is in the British Library Department of Manuscripts (Add MSS 31847).

The Chillingham toad was described by Lord Ravensworth and by J. Raine (*Archaelogia Aeliana* NS 3 [1858], 1–8 and 277–287 respectively). The Blois toad was further described by A. de Rochas (*La Nature* No. 606 [1885], 85–87). Two valuable later reviews are those by P. H. Gosse (*Romance of Natural History* 2nd Ser; London 1861, 147–190) and by Sir Ray Lankester (*Diversions of a Naturalist;* London 1915, 376–382). A Russian review by Dr. I. Tarnani was in the St. Petersburg *Revue des Sciences Naturelles* (2 [1891], 225–231). Charles Dawson described the Brighton toad in the *Annual Report of the Brighton and Hove Natural History and Philosophical Society* (1901, 1–7); there was also an article about it in the *Illustrated London News* (20 April 1901).

Like many other American celebrities, Old Rip is the subject of an admiring biography: *The Story of Old Rip* by H. V. O'Brien (Eastland 1965). The Fortean interest in the toad in the hole is discussed in *Living Wonders* by J. Michell and R. J. M. Rickard (London 1982, 98–102). Two excellent accounts of the toad in the hole, with extensive lists of references, are those by Bob Skinner (*Toad in the Hole;* London 1986) and W. R. Corliss (*Anomalies in Geology;* Glen Arm 1989, 60–71). Mr. Bob Skinner, Farnham, who is working on a longer treatise on this subject, and Dr. John Cooper, of the Booth Museum of Natural History, Brighton, are thanked for valuable help.